T0093340

Terricolous Lichens in India

Cladonia subradiata taken at 3,170 m, at Gangotri town, in Uttarkashi district of Uttarakhand, India on 28 Oct 2010. The sample is preserved in the CSIR-NBRI herbarium with collection No: 10-0014526 (LWG), by Himanshu Rai and Pramod Nag, using Fujifilm FinePix S5800 S800 camera.

Himanshu Rai • Dalip K. Upreti
Editors

Terricolous Lichens in India

Volume 1: Diversity Patterns and Distribution Ecology

 Springer

Editors
Dr. Himanshu Rai
National Botanical Research Institute
Council for Scientific and Industrial Research
Lucknow
Uttar Pradesh
India

Dr. Dalip K. Upreti
National Botanical Research Institute
Council for Scientific and Industrial Research
Lucknow
Uttar Pradesh
India

ISBN 978-1-4614-8735-7   ISBN 978-1-4614-8736-4 (eBook)
DOI 10.1007/978-1-4614-8736-4
Springer New York Heidelberg Dordrecht London

Library of Congress Control Number: 2013950764

Printed on acid-free paper

Springer is part of Springer Science+Business Media (www.springer.com)

Preface

An organism occurs in a characteristic, limited range of habitats and within this range they are found to be most abundant indicating their specific environmental optimum (Körner 2003). The distribution of organisms is strongly influenced by factors such as elevation, precipitation, moisture, temperature, and nutrients in the substratum (Huang 2010). In last 50 years though considerable attention has been accorded in documenting the taxonomic diversity of lichens in India, investigations of their community ecology have only recently begun, and those so far undertaken, except some instances (Rai et al. 2011, 2012) have not explored the distribution ecology of terricolous lichens as a functional group (Negi 2000; Negi and Upreti 2000; Pinokiyo et al. 2008). With increase in understanding on soil crust lichens, their functional role in maintenance of physical stability, hydrology, and nutrient pool of soil crust is well recognized worldwide (Elbert et al. 2012). The investigations on Indian terricolous lichens were initiated at lichenology laboratory of CSIR-National Botanical Research Institute (NBRI), as an assessment of their diversity in Western Himalaya and their role in soil stabilization in alpine habitats (Rai 2012). The study revealed a substantial diversity of terricolous lichens and found that soil lichens play a very crucial role in the stabilization of soil crust, soil respiration, amelioration of soil temperature, and growth of soil microflora. In the course of study, various patterns and factors of terricolous lichen diversity were observed and the need for a publication dealing with these aspects was realized, leading to conceptualization of this volume.

The Vol. 1 of *Terricolous Lichens in India*, in five chapters discusses lichenology in India with special reference to terricolous lichens (Chap. 1); comparative assessment of biological soil crusts (BSC) development in India with global patterns (Chap. 2); altitudinal patterns of soil crust lichens in India using generalized additive models (*GAM*; Chap. 3); role of novel molecular clades of *Asterochloris* in geographical distribution patterns of *Cladonia*—a dominant soil crust lichen (Chap. 4) and photobiont diversity of soil lichens along substrate ecology and altitudinal gradients in Himalayas (Chap. 5). The volume enumerates various patterns and factors of terricolous lichen diversity in India, as a prelude to Vol. 2 which deals with the taxonomy of Indian soil crust lichens. The book should be of interest to the specialists and also intends to generate interest among ecologists, biologists, naturalists, teachers, students, protected area managers, policy makers, and conservation agencies. We hope that this book will widen the overall understanding of Indian lichens and specifically the terricolous lichens, both for native as well as international workers and would serve as foundation of many more taxonomic as well as applied researches in Indian lichens.

References

Elbert W, Weber B, Burrows S, Steinkamp J, Büdel B, Andreae MO, Pöschl U (2012) Contribution of cryptogamic covers to the global cycles of carbon and nitrogen. Nat Geosci 5:459–462

Huang M (2010) Altitudinal patterns of *Stereocaulon* (Lichenized Ascomycota) in China. Acta Oecol 36:173–178

Körner C (2003) Alpine plant life—functional plant ecology of high mountain ecosystems, 2nd edn. Springer, Heidelberg, p 344

Negi HR (2000) On the patterns of abundance and diversity of macrolichens of Chopta–Tungnath in Garhwal Himalaya. J Biosci 25:367–378

Negi HR, Upreti DK (2000) Species diversity and relative abundance of lichens in Rumbak catchment of Hemis National Park in Ladakh. Curr Sci 78:1105–1112

Pinokiyo A, Singh KP, Singh JS (2008) Diversity and distribution of lichens in relation to altitude within a protected biodiversity hot spot, north-east India. Lichenologist 40:47–62

Rai H (2012) Studies on diversity of terricolous lichens of Garhwal Himalaya with special reference to their role in soil stability. PhD Thesis. H.N.B Garhwal University. Srinagar (Garhwal), Uttarakhand, India

Rai H, Khare R, Gupta RK, Upreti DK (2011) Terricolous lichens as indicator of anthropogenic disturbances in a high altitude grassland in Garhwal (Western Himalaya), India. Bot Oriental 8:16–23

Rai H, Upreti DK, Gupta RK (2012) Diversity and distribution of terricolous lichens as indicator of habitat heterogeneity and grazing induced trampling in a temperate-alpine shrub and meadow. Biodivers Conserv 21:97–113

CSIR-NBRI, Lucknow, U.P. India   Himanshu Rai
July 2013   Dalip Kumar Upreti

Acknowledgements

We are thankful to Dr. Roger Rosentreter (BLM, Idaho, U.S.A), Dr. Jayne Belnap (Moab, Utah, U.S.A), and Dr. Christoph Scheidegger (Swiss Federal Research Institute, WSL, Birmensdorf) for their continuous support and advice during the realization of this publication. We are also thankful to Dr. Pradeep K. Divakar (Department of Plant Biology II, College of Pharmacy, Complutense University, Madrid, Spain) for his time to time help with literature pertaining to various soil lichen genera. We are thankful to Dr. G. P. Sinha (Scientist, Central regional circle, Allahabad) for the crucial help with lichen samples of Sikkim. Further, we are grateful to all our contributors for the cooperation they extended throughout the project. Finally, we would like to thank the editorial team of Springer—Eric Stannard, Daniel Dominguez, and Andy Kwan for their patience and understanding during this project.

Contents

Contributors

Voytsekhovich Anna Department of Lichenology and Bryology, M.H. Kholodny Institute of Botany, National Academy of Sciences of Ukraine, Kyiv, Ukraine

Chitra Bahadur Baniya Central Department of Botany, Tribhuvan University, Kirtipur, Kathmandu, Nepal

Roshni Khare Lichenology laboratory, Plant Diversity, Systematics and Herbarium Division, CSIR-National Botanical Research Institute, Lucknow, Uttar Pradesh, INDIA

Dymytrova Lyudmyla Department of Lichenology and Bryology, M.H. Kholodny Institute of Botany, National Academy of Sciences of Ukraine, Kyiv, Ukraine

Ondřej Peksa The West Bohemian Museum in Pilsen, Plzeň, Czech Republic

Himanshu Rai Lichenology laboratory, Plant Diversity, Systematics and Herbarium Division, CSIR-National Botanical Research Institute, Lucknow, Uttar Pradesh, INDIA

Tereza Řídká Department of Botany, Faculty of Science, Charles University, Prague, Czech Republic

Roger Rosentreter Idaho Bureau of Land Management, Boise, ID, USA

Pavel Škaloud Department of Botany, Faculty of Science, Charles University, Prague, Czech Republic

Dalip Kumar Upreti Lichenology laboratory, Plant Diversity, Systematics and Herbarium Division, CSIR-National Botanical Research Institute, Lucknow, Uttar Pradesh, INDIA

Chapter 1
Lichenological Studies in India with Reference to Terricolous Lichens

Himanshu Rai, Roshni Khare and Dalip Kumar Upreti

1 Introduction

Lichens, a mutualistic association of a dominant fungus (mycobiont) and a green (phycobiont) and/or blue-green algae (cyanobiont), are by far known as one of the most successful symbionts in nature (Galloway 1992). At times, strictly regarded as an ecological instead of a systematic group, lichens have developed a specialized mode of nutrition, where an algal/blue-green algal partner is the sole source of carbohydrate for the fugal partner, which envelops their photosynthetic partner forming a discreet thallus (Hawksworth 1988). Thus lichen thallus is a relatively stable and well-balanced symbiotic system with both heterotrophic and autotrophic components and can be aptly regarded as a self-contained miniature ecosystem (Farrar 1976; Seaward 1988). About one-fifth of all known extant fungal species form obligate mutualistic symbiotic associations with green algae and/or cyanobacteria. Presently, the consensus of known lichenized taxa amounts to 13,500 (Hawksworth et al. 1983; Hawksworth and Hill 1984; Hawksworth 1991) revised from the earlier estimates of between 15,000 and 20,000 (Galloway 1992), constituting about 23 genera of green and 15 genera of blue-green algae (Purvis 2000).

Of all known lichenized fungi, 98 % are ascomycetes, 1.6 % deuteromycetes, and 0.4 % basidiomycetes (Honegger 2008). Nearly 40 genera of algae and cyanobacteria have been reported as photobionts in lichens (Tschermak-Woess 1988; Büdel 1992; Friedl and Büdel 2008; Honegger 2009). About 85 % of lichen-forming fungi associate with green algae (often referred to as chlorolichens; Ahmadjian 1993), about 10 % with cyanobacteria (cyanolichens; Ahmadjian 1993), and about 4 %,

H. Rai (✉) · R. Khare · D. K. Upreti
Lichenology laboratory, Plant Diversity, Systematics and Herbarium Division
CSIR-National Botanical Research Institute,
Rana Pratap Marg, Lucknow
Uttar Pradesh-226001, INDIA
e-mail: himanshurai08@yahoo.com

H. Rai, D. K. Upreti (eds.), *Terricolous Lichens in India,*
DOI 10.1007/978-1-4614-8736-4_1, © Springer Science+Business Media New York 2014

the so-called cephalodiate species, with both, simultaneously (Friedl and Büdel 2008). Three genera, *Trebouxia*, *Trentepohlia*, and *Nostoc*, are the most frequent photobionts in lichens. The genera *Trebouxia* and *Trentepohlia* are of eukaryotic nature and belong to the green algae; the genus *Nostoc* belongs to the oxygenic photosynthetic bacteria (cyanobacteria; Friedl and Büdel 2008). Among the cyano-bacteria-containing lichen species, 10% are bipartite (having cyanobacteria as the only photosynthetic partner) and 3–4% are tripartite (having two photosynthetic partners—a green algae and a cyanobacteria; Rai and Bergman 2002). In tripartite cyanolichens, cyanobacteria is restricted to a special gall like external or internal structure called cephalodia, in which the fungal partner creates microaerobic conditions to facilitate cyanobacterial nitrogen fixation (Honegger 2001; Rikkinen 2009).

The unique physiological (i.e., poikilohydric metabolism, secondary metabolite production, production of antifreeze, and UV masking agents) and anatomical (i.e., absence of waxy cuticle, absence of root, and absorption of water and nutrients passively from the environment) peculiarities make lichens some of the most tolerant as well as sensitive organisms on the planet. Constrains and vulnerabilities of lichen anatomy and physiology allow them to inhabit all sorts of habitats in major terrestrial biomes of earth, utilizing all available substratum (i.e., soil, rocks, barks, and manmade surfaces like plasters etc.; Galloway 1992).

The poikilohydrous nature of lichen thalli, which passively increase or decrease water status according to atmospheric humidity, helps them in rapid activation of their metabolism, especially photosynthetic apparatus in case of recovery from winter dormancy or desiccation (Nash 1996; Kappen 2000). For survival in harsh habitats, lichens have evolved strategies such as right choice of photobiont, water-holding structures, and tolerance to osmotic stress (Rundel 1998; Richardson 2002; Oksanen 2006). While phycolichens (green algal) are able to activate their photosynthesis with water vapor, cyanobacteria in cyanolichens need liquid water, that is why phy-colichens are better survivors in dryer habitats than cyanolichens, which are largely (50%) represented in humid tropics (Rundel 1998; Richardson 2002; Oksanen 2006). Some cyanolichens with gelatinous polysaccharides-containing thalli and phycoli-chens with cushiony water-storing thalli are able to extend their daily metabolism compared with thin, easily drying lichen species (Richardson 2002; Oksanen 2006).

Lichens exhibit a wide range of habitat differentiation. Although a shift of sub-stratum is seen in lichens due to climate change and rapidly changing land-use pat-terns, lichen genera mostly prefer specific type of substratum (Tretiach and Brown 1995). The form of lichen vegetation depends largely upon shape, structure, water relations, and the chemistry of the substrates. On the basis of substrates, the lichens can be classified into habitat subsets—saxicolous (inhabiting rocks and stones), corticolous (growing on tree barks), terricolous (soil inhabiting), ramicolous (grow-ing on twigs), muscicolous (growing over mosses), and omnicolous (inhabiting various substrates and manmade structures; Fig. 1.1). Among these habitat subsets, epiphytic lichens (corticolous) and soil lichens (terricolous lichens) are excellent indicators of ecosystem quality (Will-Wolf 2002).

Fig. 1.1 Habitat subset categories in lichens. **a** Saxicolous lichens (on rock). **b** Corticolous lichens (on bark). **c** Terricolous lichens (on soil). **d** Ramicolous lichens (on twig). **e** Muscicolous lichens (on mosses). **f** Omnicolous lichens (on manmade structures; here, iron railway track sleepers); inset: magnified lichen thallus

Terricolous lichens are a major constituent of cryptogamic biological soil crusts (BSCs) in arid/water delimited habitats and play important functional role in maintaining the fertility of soils through extensive contribution in the nitrogen dynamics of soil sink. Lichen-dominated BSC as a constituent of cryptogamic ground cover (CGC) along with cryptogamic plant cover (CPC) constitute a global continuum of cryptogamic cover acting as a major sink of atmospheric CO_2 and nitrogen, accounting for about 7% of net primary production and about half of biologically fixed nitrogen in terrestrial biomes (Elbert et al. 2009, 2012).

2 Lichenological Researches in India

In the last two centuries, there have been substantial advancements in lichenological studies worldwide, and applied fields such as lichen-based biomonitoring, bioprospection of lichen compounds, lichenometry, and studies on the role of lichens in ecosystem services have gained momentum. However, the Indian lichenological research though has focused largely on the taxonomical aspect of lichenized fungi; researches in other aspects (i.e., biomonitoring, bioprospection, ecology, and assessment of their functional role in ecosystem) have also started.

2.1 A Brief Historical Account[1]

Lichenological investigation in the Indian subcontinent was initiated by Linnaeus in 1753, who mentioned a single lichen species, now under the genus *Roccella* (*R. montagenii*) in his iconic publication *Species Plantarum*. In the years 1810 and 1814, Eric Acharius described four species of lichens from India. During the nineteenth and twentieth centuries, lichens from the Indian subcontinent were collected largely by European botanists, naturalists, British army personnel, and Jesuit missionaries (Fries 1825; Bêlanger 1838; Montagne 1842; Taylor 1852; Nylander 1860, 1867, 1869, 1873; Müller 1874, 1891, 1892, 1895; Hue 1898, 1899, 1900a, b, 1901; Jatta 1902, 1905, 1911; Smith 1931) and about 1,000 species of lichens from all over the country were described (Upreti 2001b). Quaraishi (1928) was probably the first Indian who recorded 35 species of lichens near Mussoorie in western Himalayas. Later, Dr. S. R. Kashyap initiated the collection of Indian lichens which were determined by Dr. A. L. Smith and the data were published in the form of a hand book (Chopra 1934). The more systematic study on Indian lichens was carried out by Late Dr. D. D. Awasthi (1922–2011), in the fifties of the last century. Studies on Indian lichens were started by Dr. D. D. Awasthi in the early 1950s at the Department of Botany, Lucknow University, Lucknow. This center remained active for about 40 years and

---

[1] For a detailed historical account of lichenological researches (mainly taxonomic) in India, see Upreti 2001b and Singh 2011.

considerable progress, particularly in respect of the generic, floristic, and monographic studies on lichens of India, as well as Nepal, was made (Upreti 2001b). The work by Dr. Awasthi at Lucknow University resulted in about 80 publications (Singh 2011). His consolidated account of lichens of India in the form of publications such as *A Handbook of Lichens* (Awasthi 2000a), *Lichenology in Indian Subcontinent—A Supplement to A Handbook of Lichens* (Awasthi 2000b), where he listed over 2,000 species, *Compendium of Macrolichens from India, Nepal and Sri Lanka*, where he dealt with about 970 species (Awasthi 2007), are regarded as the backbone of taxonomy of Indian lichens and serve as baseline data on Indian lichens.

The institutions whose consistent efforts in researches on Indian lichens played instrumental role in present development of Indian lichenology are Lichenology laboratory and herbarium (LWG), National Botanical Research Institute (NBRI), Lucknow; Botany department, Lucknow University (LWU), Lucknow (inactive centre, mainly consisted of collections of Dr. D.D. Awasthi, now lodged at NBRI) ; Agharkar Research Institute (ARI), Pune; Botanical Survey (BSI) of India (north eastern and central circle) and M.S. Swaminathan Research Foundation (MSSRF), Chennai. Among these lichenological centers, NBRI holds the largest collection of Indian lichens from all over the country, and along with taxonomic researches (Divakar and Upreti 2005; Upreti 1985a, b, 1987, 1988, 1990, 1991a, b, 1992, 1993a, b, 1994, 1997b, c), it has also initiated researches in other fields of lichenology such as ethnolichenological aspects (Saklani and Upreti 1992; Upreti 1996, 2001a, b, c; Upreti and Negi 1996; Kumar and Upreti 2001; Upreti et al. 2005), antimicrobial activity of lichen compounds (Tiwari et al. 2011a, b), pollution monitoring (Upreti and Pandey 1994, 1999; Nayaka et al. 2003; Satya and Upreti 2009; Shukla and Upreti 2007a, b, 2009, 2012[2]), conservation (Upreti 1995; Upreti and Nayaka 2008), lichenometry (Joshi and Upreti 2010), climate-warming studies using lichen-based passive temperature-enhancing devices (Rai et al. 2010), and biomonitoring and ecophysiological studies of specific functional groups with reference to zooanthropogenic pressures and macroscale climatic variables (i.e., terricolous lichens and cyanolichens; Khare et al. 2013; Rai et al. 2011, 2012, 2013a, b). The revisionary monograph by NBRI center on parmelioid lichens, *Parmelioid Lichens in India: A Revisionary Study*, serves as the main reference publication on Indian Parmeliaceae (Divakar and Upreti 2005). ARI, though mainly involved in taxonomical researches of Western Ghats (Pyrenocarpous, Graphidaceous, and Thelotremataceous lichens), has worked extensively on in vitro culture methodologies of lichens, antioxidants, and pharmaceutically important activities of lichen compounds (Behera and Makhija 2001, 2002; Dubey and Makhija 2008, 2010; Makhija and Patwardhan 1987, 1993, 1995, 1997, 1998a, b, c; Makhija and Adawadkar 2001, 2002, 2007; Makhija et al. 2005, 2009; Behera et al. 2000, 2004, 2005a, b, 2006a, b, c, 2009; Gaikwad et al. 2012; Verma et al. 2008, 2011a, b, 2012a, b).BSI, being a government agency with mandate of surveying the plant resources of the India, has exclusively done taxonomic studies of Indian lichens, mainly of the north-eastern states, Sikkim and Darjeeling regions of the country

---

[2] The chapter reviews pollution monitoring studies in India.

(Jagdeesh Ram et al. 2005, 2007, 2009, 2012; Jagdeesh Ram and Sinha 2009a, b, 2010; Singh 1977, 1979, 1980, 2011; Singh and Sinha 2010). The recently published *Lichens: An Annotated Checklist*, serves as the latest record of lichen diversity in India (Singh and Sinha 2010). MSSRF is involved in investigations on lichen diversity of the east coast (Mohan and Hariharan 1999; Balaji and Hariharan 2004, 2005), bioprospecting (antimicrobial activity) of lichen secondary compounds (Balaji et al. 2006; Balaji and Hariharan 2007), and molecular and biotechnological aspects (Valarmathi and Hariharan 2007; Valarmathi et al. 2009).

3 Lichenological Research in India with Reference to Terricolous Lichens

Terricolous lichens as a group have not been dealt separately and are mentioned only in taxonomic descriptions and enumeration of specific regions. Lichenological investigations in India with reference to terricolous lichens can be described in three major heads: (i) taxonomic records and enumerations, (ii) ethnopharmacological studies, and (iii) functional ecology and bioindicator studies.

3.1 Taxonomic Records and Enumerations

Awasthi (1965) catalogued the Indian lichens and reported taxa of soil lichen known till then. Awasthi and Awasthi (1985) described soil lichen genera *Alectoria, Bryoria*, and *Sulacria* from India and Nepal. Upreti (1985a), in his studies on *Baeomyces*, described four species (*B. fungoides, B. pachypus, B. roseus, B. sorediifer*) from Darjeeling, eastern Himalayas, and Palni Hills, South India. Upreti (1985b) reported soil lichen genus *Cladia aggregata* from India and Nepal. Awasthi (1988, 1991) keyed out macro- and microlichens of India, Nepal, and Sri Lanka and mentioned the occurrence of many species of terricolous lichens, i.e., *Collema, Endocarpon, Fulgensia, Solorina, Squamaria*, and *Toninia sediflora*. Upreti (1987) provided key of 62 species of *Cladonia* from Nepal and India. Upreti and Büdel (1990) reported soil lichen species *Heppia lutosa* (Ach.) Nyl. from northern India. Pant and Upreti (1993) described terricolous species *Diploschistes diacapsis, D. muscorum, D. nepalensis*, and *D. scruposus* from India and Nepal. Pant and Upreti (1999) gave detailed taxonomic description along with key of terricolous lichen genus *Stereocaulon* in India and Nepal. Ahti and Upreti (2004) described two new species of terricolous genus *Cladonia* (*Cladonia awasthiana* and *C. indica*) from Himalayas. Khare et al. (2009) described taxonomic diversity of soil lichens in India. Upreti et al. (2010) reported new records of squamulose soil lichens *Lecidoma demissum, Psora decipiens, P. himalayana, Toninia cinereovirens*, and *T. tristis* from western Himalayas. Upreti and Divakar (2010) reported a new species *Sticta indica* in soil over rock, from Rudraprayag, Uttarakhand in the western Himalayas. McCune et al. (2012), in their revision of *Hypogymnia* from Himalaya of India and Nepal, described terricolous species *Hypogymnia alpina*, *H. bitteri*, and, *H. vittata* from Himalayas.

Upreti and Negi (1995) mentioned terricolous lichens species of *Cladonia* and *Stereocaulon* from the temperate alpine habitats (2,700–4,500 m) of the northwestern part of Nandadevi Biosphere Reserve, Chamoli, in the Western Himalayas. Upreti and Negi (1998) described occurrence of terricolous lichen taxa *Lobaria*, *Peltigera*, *Cladonia*, *Stereocaulon*, *Umbilicaria*, *Rhizoplaca*, *Cetraria*, *Hypogymnia*, *Parmelia*, *Ramalina*, *Usnea*, *Caloplaca*, *Heterodermia*, and *Phaeophyscia* in Chopta, Tungnath. Upreti and Nayaka (2000) described occurrence of soil lichen species of genera *Acarospora*, *Cladonia*, *Coccocarpia*, *Collema*, *Leptogium*, *Rhizoplaca*, *Nephroma*, *Bulbothrix*, *Cetraria*, *Everniastrum*, *Flavoparmelia*, *Flavopuctelia*, *Hypogymnia*, *Hypotrachyna*, *Melanelia*, *Nephromopsis*, *Parmelia*, *Peltigera*, *Heterodermia*, *Phaeophyscia*, *Ramalina*, *Stereocaulon*, *Lobaria*, *Sticta*, *Bryoria*, and *Diploschistes* from Himachal Pradesh. Upreti et al. (2001) describe soil lichens in Askote Sandev Botanical Hotspot of Pithoragarh district, Uttarakhand. Upreti et al. (2002) described species of terricolous genera *Cladonia*, *Ramalina*, *Lobaria*, *Leptogium*, *Heterodermia*, *Phaeophyscia*, *Prmotrema*, *Bulbothrix*, *Flavoparmelia*, *Diploschistes* from Sirmaur, Himachal Pradesh. Hariharan et al. (2003) mentioned *Cladonia*, *Leptogium*, *Everniastrum*, *Flavoparmelia*, *Parmellinella*, *Parmotrema*, *Heterodermia*, *Sticta* from Shevaroy hills of Eastern Ghats, India. Rout et al. (2004) described *Cladonia furcata*, *Lobaria kurokawa*, and *Sticta nylanderiana* in enumeration of lichens of Sessa Orchid sanctuary, West Kameng, Arunachal Pradesh. Upreti et al. (2004) mentioned many soil lichens, e.g., *Cladonia*, *Melanelia*, and *Peltigera* from Gangotri and Gomukh areas of Uttaranachal, India. Sheik et al. (2006) described terricolous lichen species, *Candellariella aurella*, *Catapyrenium cinereum Cladonia awasthiana*, *C. fimbriana*, *C. ochrochlora*, *C. pocillum*, *C. pyxidata*, *Collema limosum*, *Lecanora himalayae*, *Lecidella alaiensis*, *L. euphoria*, *L. caesioatra*, *L. flavosorediata*, *Peltigera canina*, *P. dolichorrhiza*, *P. horizontalis*, *P. polydactyla*, *P. rufescence*, *Physconia muscigena*, *P. pulverulenta var. argyphaea*, *Solorina bispora*, *Squamarina cartilaginea*, *Stereocaulon glareosum*, *Toninia coeruleonigricans*, *Xanthoparmelia somlensis*, and *X. taractica* from Jammu and Kashmir. Dubey et al. (2007) described occurrence of soil lichen *Cladonia corymbescens* from Along town, West Siang district, Arunachal Pradesh. Joshi et al. (2007) described occurrence of soil lichen *Stereocaulon foliolosum* from Khaliya top and Kalamuni in Pithoragarh district of Uttarakhand. Rawat et al. (2009) described soil lichens *Biatora*, *Cladonia squamosa*, and *Lobaria retigera* from Mandal and adjoining localities in Chamoli district of Uttarakhand.

3.2 Ethnopharmacological Studies

Subramanian and Ramakrishnan (1964) reported occurrence of five essential amino acids from terricolous lichen *Peltigera canina*. Saklani and Upreti (1992) while reporting folk uses of lichens from Sikkim reported that *Peltigera canina* is used as a remedy for liver ailments and that its high content of amino acid methionine may contribute to its alleged curative powers. Upreti and Negi (1996) reported ethnobotanical utilization of *Thamnolia vermicularis* (Swartz.) Ach. in Schaerer, and found that this terricolous lichen is used for killing worms in milk by indigenous ethnic tribes, *Bhotias*, and was also used for ritual offerings in religious prayers. Upreti (1996) described that terricolous lichen *Heterodermia diademata* is used on wounds as protection from infection and water. Shahi et al. (2000) described successful use of aqueous extract of soil lichen *Everniastrum cirrhatum* for curing superficial fungal infections in humans. Upreti (2001a) discussed exploitation of soil lichens *Bulbothrix meizospora*, *Everniastrum cirrhatum*, *E. nepalense*, *Heterodermia diademata*, and *Leptogium askotense* from Pithoragarh, Uttarakhand. Upreti (2001b) mentioned that taxonomy of soil lichen genus *Bryoria*, *Buellia*, *Cladonia*, *Heterodermia*, *Leptogium*, *Peltigera*, *Stereocaulon*, *Sticta*, and *Ramalina* has been well worked out and described the ethnic uses of terricolous lichen species of *Everniastrum*, *Heterodermia*, *Peltigera*, *Stereocaulon*, *Thamnolia vermicularis*, *Cetraria*, *Lobaria*, and *Ramalina*. Kumar and Upreti (2001) described medicinal value of terricolous lichen species *Parmelia sulcata* as described in ancient Indian medicinal system. Upreti et al. (2005) discussed the commercial exploitation of soil lichens *Everniastrum cirrhatum*, *Heterodermia*, *Cladonia*, and *Thamnolia vermicularis*. Upreti (2001c) discussed utilization of some soil lichens as human food and condiments. Shukla et al. (2003) described antifungal activity of soil lichen *Stereocaulon* against some storage fungi. Gupta et al. (2007) reported antimycobacterial activity of some terricolous lichens (*Leptogium pedicellatum*, *Lobaria isidiosa*, *Stereocaulon foliolosum*) from Kumaun Himalayas. Kumar et al. (2010) reported antibacterial activity of cyanolichen—*Collema auriformis* from Kolli hills, Tamil Nadu.

3.3   Functional Ecology and Bioindicator Studies

Upreti (1997a) described the distribution of terricolous lichen taxa *Cladonia* and *Stereocaulon* along gradients of altitude and substrate along with other Himalayan lichens. Upreti (1995) described factors responsible for loss of diversity of terricolous lichen species of *Heterodermia*, *Peltigera*, *Stereocaulon*, *Thamnolia*, *Diploschistes*, *Cladonia*, *Phaeophyscia*, *Parmelia*, and *Leptogium* in India. Negi and Upreti (2000) in their studies on species diversity and relative abundance of lichens in Rumbak catchment of Hemis National Park, Ladakh, described the distribution patterns of soil lichen genus *Cetraria*, *Cladonia*, *Lecanora*, and *Physcia*, with relation to other substrates and discussed unstable top layer and trampling caused by grazing as the main cause of poor turnout on soil than on other substratum like rocks. Nayaka et al. (2009) assessed the photosynthetic vitality of the soil lichen *Cladonia subconistea* from Himalaya.

Rai et al. (2011, 2012), in their studies in an alpine shrub and meadow in Himalayas, found that the diversity of terricolous lichens was delimited both by grazing pressures and decrease in soil cover along with increasing altitudinal gradient. The study established that fruticose (*Stereocaulon* spp) and dimorphic/compound[3] (*Cladonia* spp.) were the dominant growth forms owing to their low palatability and better tolerance of the harsh climate at higher altitudes. The distribution of grazing and harsh climate-tolerant growth forms is a common feature shared by terricolous lichens worldwide (Chap. 2).

Terricolous lichens have been found instrumental in maintaining higher carbon(C)–nitrogen(N) content[4] of the soil crust (Rai et al. 2013b). Studies in alpine habitats have shown that though the C–N content of terricolous lichens with cyanobionts was higher than that of chlorolichens, C–N content of terricolous tripartite cyanolichen, *Stereocaulon foliolosum* (N—0.862 ± 0.01; C—37.3 ± 0.04) was greater than that of monopartite cyanolichen species, *Peltigera praetextata* (N—0.562 ± 0.02; C—26.42 ± 0.02) and *P. rufescens* (N—0.498 ± 0.01; C—32.02 ± 0.01; Rai et al. 2013b). Terricolous soil lichens form a closed feedback loop with reference to C–N dynamics where the elements (C, N) hold up in the thallus, enrich the substratum by leaching of metabolites from, and decomposition of the thallus, which are subsequently fixed by cyanolichens (Rai et al. 2013b).

Rai et al. (2013a), in their studies on control of photobionts on diversity and distribution of terricolous lichens, found that the distribution of chlorolichens and cyanolichens was along the gradients of precipitation and altitudes, where cyanolichens preferred habitats with higher precipitation as they need liquid water for their active physiology, whereas chlorolichens were widely distributed as their hydration needs for proper functioning of their physiology are met even with air moisture.

Rai (2012), in his studies on functional role of terricolous lichens in Garhwal Himalayas, found that terricolous lichens play a key role in amelioration of a number of physicochemical properties of soil (i.e., aggregate stability, soil temperature, soil microbial respiration and, soil carbon–nitrogen content), as they are habitat specialists and usually inhabit habitats with poor/unstable nutrient content, highly acidic pH and harsh climate characterized by longer periods of subzero minimum temperature, high wind velocity, fluctuating relative humidity, and hydrological regimes.

Rai (2012), employing generalized additive models (GAM) on elevational patterns of terricolous lichens in Garhwal Himalayas, concluded that maximum terricolous lichen species' richness is at mid-altitudes (3,000–3,100 m), which can be attributed to decrease in competition from vascular plants, as at these heights the tree line in the Himalayas starts to thin out (Fig. 1.1). The differential peak altitudinal distribution of cyano- as well as chloroterricolous lichens can be due to different efficiency of the both groups for water utilization (Fig. 1.2). Further, the higher peak altitudinal distribution of compound and fruticose growth forms is due to the tolerant nature of these forms to harsh climate extremes and deterrence to grazing in Garhwal Himalayas (Fig. 1.3). Analyzing the GAM plots of different dominant families of terricolous lichens in Garhwal Himalayas, Rai (2012) concluded that though some families inhabit up to 3,900 m altitude (*Stereocaulaceae* and *Parmeli-*

---

[3] Squamules as primary thallus bearing erect fruticose body as secondary thallus.

[4] Values measured as percent dry weight, determined using CHNS-O Elementar (Vario EL III) using sulfanilic acid (SIGMA-ALDRICH™) as standard.

Fig. 1.2 Relationship between elevation and terricolous lichen species richness from Garhwal Himalaya. **a** Total terricolous lichen species richness. **b** Cyanolichen species richness. **c** Chlorolichen species richness. The *fitted regression line* represents the statistically significant P (≤ 0.001) smooth spline(s) after using GAM with approximately four degrees of freedom

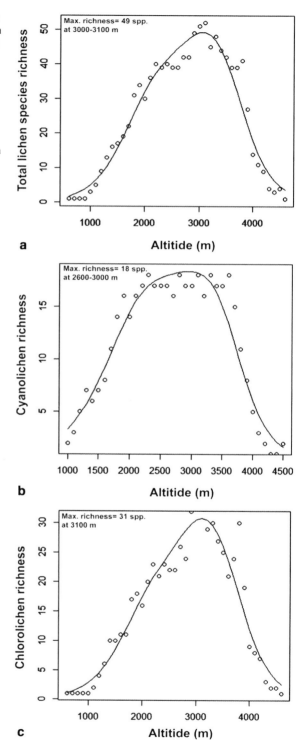

Fig. 1.3 Relationship between elevation and terricolous lichen species richness from Garhwal Himalaya. **a** Foliose terricolous lichen species richness. **b** Fruticose terricolous lichen species richness. **c** Dimorphic/compound terricolous lichen species richness. The *fitted regression line* represents the statistically significant *P* (≤0.001) smooth spline(s) after using GAM with approximately four degrees of freedom

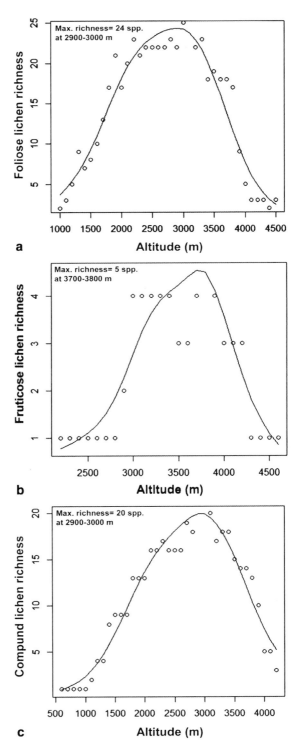

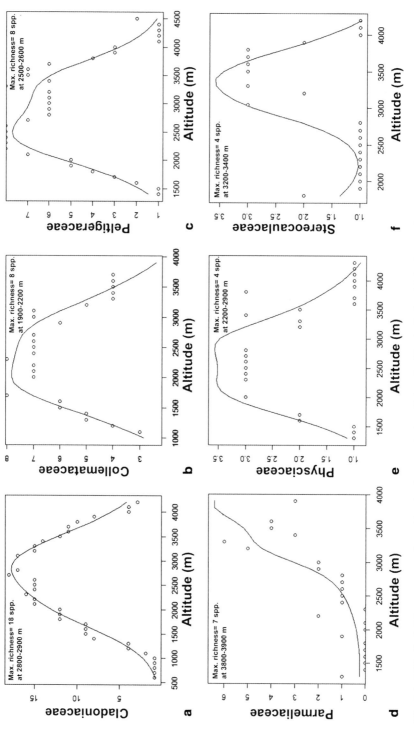

Fig. 1.4 Altitudinal richness pattern showed by dominant terricolous lichen families in Garhwal Himalaya. **a** Cladoniaceae. **b** Collemataceae. **c** Peltigeraceae. **d** Parmeliaceae. **e** Physciaceae. **f** Stereocaulaceae. The *fitted regression line* represents the statistically significant P (≤ 0.001) smooth spline(s) after using GAM with approximately four degrees of freedom

aceae), the maximum species' richness of *Cladoniaceae*, at optimum mid-altitudes (about 3,000 m) in Garhwal Himalaya is due to the tolerant growth forms of the species in the family (e.g., *Cladia* and *Cladonia*; Fig. 1.4).

4 Conclusion

Although lichenological researches in India are mainly focused on taxonomy of lichens, studies on other lichenological aspects are gaining momentum (Upreti 2001b). Lichens are now being studied for the bioprospection of secondary lichen compounds, biomonitoring and functional ecology. Terricolous lichens constitute about 9–14 % of total lichen flora of India (Rai et al. 2011), and due to their ability to inhabit habitats with harsh climate, poor soil nutrients, acidic pH, unstable air moisture, and soil hydration regime, they play a vital role in functional ecology of soils. Controlled by constrains of photobiont physiological efficiency terricolous lichens extend up to 3,700 m, where in absence of competitive pressures by other ground vegetation, they optimize the soil microenvironment, for soil microflora, which helps crucially for development of successional vegetations. Although there is not much work done on terricolous lichens and their role in soil functional ecology in India, the work actually done points to their pivotal role and needs further investigation.

References

Acharius E (1810) Lichenographia Universalis. Gottingen

Acharius E (1814) Synopsis Methodica Lichenum. Lund

Ahmadjian V (1993) The lichen symbiosis. Wiley, New York

Ahti T, Upreti DK (2004) Two new species of *Cladonia* from the Himalayas. Bibl Lichenol 88:9–13

Awasthi DD (1988) A key to the Macrolichens of India and Nepal. J Hattori Bot Lab 65:207–302

Awasthi DD (1991) A key to the microlichens of India, Nepal & Sri Lanka. Bibl Lichenol 40:1–336

Awasthi DD (2000a) Lichenology in Indian subcontinent: a supplement to a handbook of lichens. Bishen Singh Mahendralpal Singh, Dehradun

Awasthi DD (2007) A Compendium of the Macrolichens from India, Nepal and Sri Lanka. Bishen Singh Mahendrapal Singh, Dehradun

Awasthi G, Awasthi DD (1985) Lichen genera *Alectoria*, *Bryoria* and *Sulacria* from India and Nepal. Candollea 40:305–320

Awasthi DD (1965) Catalogue of Lichens from India, Nepal, Pakistan and Ceylon. Beih Nova Hedw 17:1–137

Awasthl DD (2000b) A handbook of Lichens. Bishen Singh Mahendrapal Singh, Dehradun

Balaji P, Bharath P, Satyan RS, Hariharan GN (2006) *In vitro* antimicrobial activity of *Roccella montagnei* thallus extracts. J Trop Med. Plant 7:169–173

Balaji P, Hariharan GN (2004) Lichen diversity and its distribution pattern in tropical dry evergreen forests of Gundy National Park (GNP), Chennai. Indian For 130:1154–1168

Balaji P, Hariharan GN (2005) Annonated checklist of the lichens of Chennai, Tamil Nadu, India. Phytotaxonomy 5:1–7

Balaji P, Hariharan GN (2007) *In vitro* antimicrobial activity of *Parmotrema praesollediosum* thallus extracts. Res J Bot 2:54–59

Behera BC, Adawadkar B, Makhija U (2004) Capacity of some Graphidaceous lichens to scavenge superoxide and inhibition of tyrosinase and xanthine oxidase activities. Curr Sci 87:83–87

Behera BC, Adawadkar B, Makhija U (2006a) Tissue culture of selected species of the lichen genus *Graphis* and their biological activities. Fitoterapia 77:208–215

Behera BC, Makhija U (2001) Effect of various culture conditions on growth and production of salazinic acid in *Bulbothrix setschwanensis* (lichenized ascomycetes) *in vitro*. Curr Sci 80:1424–1427

Behera BC, Makhija U (2002) Inhibition of tyrosinase and xanthine oxidase by lichen species *Bulbothrix setschwanensis*. Curr Sci 82:61–66

Behera BC, Makhija U, Adawadkar B (2000) Tissue culture of *Bulbothrix setschwanensis* (lichenized ascomycetes) in vitro. Curr Sci 78:781–783

Behera BC, Verma N, Sonone A, Makhija U (2005a) Antioxidant and antibacterial activities of lichen *Usnea ghattensis in vitro*. Biotechnol Lett 27:991–995

Behera BC, Verma N, Sonone A, Makhija U (2005b) Evaluation of antioxidant potential of the cultured mycobiont of a lichen a *Usnea ghattensis*. Phytother Res 19:58–64

Behera BC, Verma N, Sonone A, Makhija U (2006b) Determination of antioxidative potential of lichen *Usnea ghattensis in vitro*. Lebensmittel-Wissenschaft und-Technologie 39:80–85

Behera BC, Verma N, Sonone A, Makhija U (2006c) Experimental studies on growth and production of usnic acid of lichen *Usnea ghattensis in vitro*. Microbiol Res 161:232–237

Behera BC, Verma N, Sonone A, Makhija U (2009) Optimization of culture conditions for lichen *Usnea ghattensis* G. Awasthi to increase biomass and antioxidant metabolite production. Food Technol Biotechnol 47:7–12

Bélanger MC (1838) Voyage aux bides Orientales Annees 1825–1829. Botanique II. Partie. Cryptogamie par ch. Belanger et Bory de St. Vince 1834–1838. Paris. Lichens, pp 113–114

Büdel B (1992) Taxonomy of lichenized prokaryotic blue-green algae. In: W Reisser (ed) Algae and Symbioses. Biopress Limited, Bristol, pp 301–324

Chopra GL (1934) Lichens of Himalayas. Punjab University Press, Punjab

Divakar PK, Upreti DK (2005) Parmelioid Lichens in India. Bishen Singh Mahendra Pal Singh, Dehradun

Dube A, Makhija U (2008) A new species of *Parmeliella* (*Pannariaceae*) from India. Lichenologist 40:209–212

Dube A, Makhija U (2010) Occurrence of four additional non-hairy species of *Leptogium* from Maharashtra, India. Lichenologist 42:701–710

Dubey U, Upreti DK, Rout J (2007) Lichen flora of Along town, West Siang district, Arunachal Pradesh. Phytotaxonomy 101:25–27

Elbert W, Weber B, Büdel B, Andreae MO, Pöschl U (2009) Microbiotic crusts on soil, rock and plants: neglected major players in the globabl cycles of caron and nitrogen. Biogeosciences Discuss 6:6983–7015

Elbert W, Weber B, Burrows S, Steinkamp J, Büdel B, Andreae MO, Pöschl U (2012) Contribution of cryptogamic covers to the global cycles of carbon and nitrogen. Nat Geosci 5:459–462

Farrar JF (1976) The lichen as an ecosystem: observation and experiment. In: DH Brown, DL Hawksworth, RH Bailey (Eds) Lichenology: progress and problems. Academic Press, London, pp 385–406

Friedl T, Büdel B (2008) Photobionts. In: TH Nash (ed) Lichen Biology, 2nd edn. Cambridge University Press, Cambridge, pp 9–26

Fries E (1825) Systema Orbis Vegetabilis. Lund

Gaikwad S, Verma N, Sharma BO, Behera BC (2012) Growth promoting effects of some lichen metabolites on probiotic bacteria. J Food Sci Technol 1–8. doi:10.1007/s13197-012-0785-x

Galloway DJ (1992) Biodiversity: a lichenological perspective. Biodivers Conserv 1:312–323

Gupta VK, Darokar MP, Saikia D, Pal A, Fatima A, Khanuja SP (2007) Antimycobacterial activity of lichens. Pharm Biol 45:200–204

Hariharan GN, Krishnamurthy KV, Upreti DK (2003) Lichens of Shevaroy hills of Eastern Ghats, India. Phytotaxonomy 3:1–23

Hawksworth DL (1991) The fungal dimensions of biodiversity: magnitude, significance, and conservation. Mycol Res 95:641–655

Hawksworth DL (1998) The variety of fungal-algal symbioses, their evolutionary significance, and the nature of lichens. Bot J Linn Soc 96:3–20

Hawksworth DL, Hill DJ (1984) The Lichen-forming fungi. Blackie, Glasgow

Hawksworth DL, Sutton BC, Anisworth GC (1983) Anisworth & Bisby's dictionary of the fungi, 7th edn. Commonwealth Mycological Institute, Kew

Honegger R (2008) Mycobionts. In: Nash TH III (ed) Lichen biology, 2nd edn. Cambridge University Press, pp 27–39

Honegger R (2001) The symbiotic phenotype of lichen-forming ascomycetes. In: Hock B (ed), The mycota IX. Fungal associations. Springer-Verlag, Berlin, pp 165–188

Honegger R (2009) Lichen-Forming Fungi and Their Photobionts. In: Deising H (ed) Plant relationships, vol 5. Springer, Berlin, pp 307–333

Hue AM (1898) Lichens extra-europaei a pluribus collectoribus ad Museum Parisiense missi. Nouv. Arch. du Mus. Hist. Natur. de Paris 3:213–280

Hue AM (1899) Lichens extra-europaei a pluribus collectoribus ad Museum Parisiense missi. Nouv. Arch. du Mus. Hist. Natur. de Paris 4:27–220

Hue AM (1900a) Lichens extra-europaei a pluribus collectoribus ad Museum Parisiense missi. Nouv. Arch. du Mus. Hist. Natur. de Paris 4:48–122

Hue AM (1900b) Lichens récoltes à Coonoor, massif du Nilghéris Chaine des Ghattes, Inde, par M. Ch. Gray en 1893. Bull. Acad. Internat. Geogr. Bot 9:251–265

Hue AM (1901) Lichens extra-europaei a pluribus collectoribus ad Museum Parisiense missi. Nouv. Arch. du Mus. Hist. Natur. de Paris 4:21–146

Jagadeesh Ram TAM, Aptroot A, Sinha GP, Singh KP (2005) New species and new records of lichenized and non lichenized pyrenocarpous ascomycetes from the Sundarbans biosphere reserve, India. Mycotaxon 91:455–461

Jagadeesh Ram TAM, Aptroot A, Sinha GP, Singh KP (2007) A new isidiate *Megalaria* species, lichenized, lichenicolous and non-lichenized ascomycetes from India. Nova Hedwig 85:139–144

Jagadeesh Ram TAM, Sinha GP (2009a) New species and new records of *Herpothallon* (lichenized Ascomycota) from India. Mycotaxon 110:37–42

Jagadeesh Ram TAM, Sinha GP (2009b) New species of *Graphis* and *Hemithecium* (lichenized Ascomycota) from Eastern Himalaya, India. Mycotaxon 110:31–35

Jagadeesh Ram TAM, Sinha GP (2010) A new species and new records of *Pyrenula* (Pyrenulaceae) from India. Lichenologist 42:51–53

Jagadeesh Ram TAM, Sinha GP, Singh KP (2009) New species and new records of *Cryptothecia* and *Herpothallon* (Arthoniales) from India. Lichenologist 41:605–613

Jagdeesh Ram TAM, Sinha GP, Singh KP (2012) Lichen flora of Suderbans Biosphere Reserve, West Bengal. Bishen Singh Mahendra Pal Singh, Dehradun, pp 384

Jatta A (1902) Licheni esotici dell Erabario Levier eracollti nell Asia meridionale nell, Oceania. Malpighia 17:3–15

Jatta A (1905) Licheni esotici dell Erabario Levier eracollti nell Asia meridiooale nell, Oceania nell Brasile e ne Madagascar. Malpighia 19:163–165

Jatta A (1911) Lichens Asia meridionalis lecti B. Luthi in Malabar et a E. Long et W. Gollan in Himalayas, Bull. Orto.d.R. Univ. Napoli 3:309–312

Joshi S, Upreti DK (2010) Lichenometric studies in vicinity of Pindari Glacier in the Bageshwar district of Uttarakhand, India. Curr Sci 99:231–234

Joshi S, Upreti DK, Punetha N (2007) Lichen flora of Munsiyari, Khaliya top and Kalamuni in Pithoragarh district of Uttarakhand. Phytotaxonomy 7:50–55

Kappen L (2000) Some aspects of great success of lichens in Antarctica. Antarct Sci 12:314–324

Khare R, Rai H, Upreti DK, Gupta RK (2013) Lichens as indicators of habitat heterogeneity in a high altitude montaine pass (Sella pass) in Eastern Himalaya. Souvenir, UGC sponsored national conference on resource management & its sustainable use, 22–23 March 2013, Pt. L.M.S. Govt. P.G. College, Rishikesh (Dehradun), Uttarakhand, pp 17

Khare R, Upreti DK, Nayaka S, Gupta RK (2009) Diversity of Soil lichens in India. In: Gupta RK, Kumar M, Vyas D (Eds) Soil Microflora, Daya Publishing House. New Delhi, pp 64–75

Kumar K, Upreti DK (2001) *Parmelia* spp. (Lichen) – in ancient medicinal plant lore of India. Econ Bot 55:458–459

Kumar RS, Thajuddin N, Venkateswari C (2010) Antibacterial activity of cyanolichen and symbiotic cyanobacteria against some selected microorganisms. Afr J Microbiol Res 4:1408–1411

Linnaeus C (1753) Species Plantarum (Stockhlom)

Makhija U, Adawadkar B (2001) Contributions to the lichen flora of the Lakshadweep (Laccadive) Islands, India. Lichenologist 33:507–512

Makhija U, Adawadkar B (2002) The lichen genus *Sclerophyton* in India. Lichenologist 34:347–350

Makhija U, Adawadkar B (2007) Trans-septate species of *Acanthothecis* and *Fissurina* from India. Lichenologist 39:165–185

Makhija U, Chitale G, Sharma B (2009) New species and new records of *Diorygma* (Graphidaceae) from India: species with convergent exciples. Mycotaxon 109:379–392

Makhija U, Dube A, Adawadkar B, Chitale G (2005) Five trans-septate species of *Hemithecium* from India. Mycotaxon 93:365–372

Makhija U, Patwardhan PG (1987) Some new and interesting lichens from India. J Econ Taxon Bot 10:497–503

Makhija U, Patwardhan PG (1993) A contribution to our knowledge of the lichen genus *Trypethelium* (Family: *Trypetheliaceae*). J Hattori Bot Lab 73:183–219

Makhija U, Patwardhan PG (1995) The Lichen genus *Arthothelium* (Family *Arthoniaceae*) in India. J Hattori Bot Lab 78:189–235

Makhija U, Patwardhan PG (1997) A new saxicolous species of *Arthothelium*. Lichenologist 29:169–172

Makhija U, Patwardhan PG (1998a) The Lichen genus *Stirtonia* (Family *Arthoniaceae*). Mycotaxon 67:287–311

Makhija U, Patwardhan, PG (1988b) The Lichen Genus *Laurera* (Family *Trypetheliaceae*) in India. Mycotaxon 31:565–590

Makhija U, Patwardhan, PG (1988c) Materials for a Lichen Flora of the Andaman Islands –IV Pyrenocarpous lichens. Mycotaxon 31:467–481

McCune B, Divakar PK, Upreti DK (2012) *Hypogymnia* in the Himalayas of India and Nepal. Lichenologist 44:595–609

Mohan MS, Hariharan GN (1999) Lichen distribution pattern in Pichavaram—a preliminary study to indicate forest disturbance in the Mangroves of South India: In: Mukerji KG et al (Eds) Biology of lichens. Aravali books international, New Delhi, pp 283–296

Montagne C (1842) Cryptogamae Nilgherrenses seu Plantarum in montibus peninsulae indicae Neelgheries dictis a cl. Perrottet collectarum enumeratio. Ann Sci Nat Bt II. Lichens (17):17–21

Müller AJ (1874–1891) Lihenologische Beitrege, 1–25, in Fragmems at intervals, Flora vols. 57–74

Müller AJ (1892) Lichenes Manipuriensis, a cl. Dr. G. Watt circa Manipur ad limites orientales Indiae Orientalis 1881–1882 lecti. J Linn Soc, Bot 29:217–231

Müller AJ (1895) Lichens Sikkimensis. Bull Herb Boissier 3:194–195

Nash TH III (1996) Photosynthesis, respiration, productivity and growth. In: Nash TH III (ed) Lichen biology. Cambridge University Presss, Press, pp 88–120

Nayak S, Upreti DK, Gadgil M, Pandey V (2003) Distribution pattern and heavy metal accumulation in lichens of Bangalore city with special reference to Lalbagh garden. Curr Sci 84:674–680

Nayaka S, Ranjan S, Saxena P, Pathre UV, Upreti DK, Singh R (2009) Assessing the vitality of the Himalayan lichens by measuring their photosynthetic performance by using chlorophyll fluorescence technique. Curr Sci 97:538–545

Negi HR, Upreti DK (2000) Species diversity and relative abundance of lichens in Rumbak catchment of Hemis National Park in Ladakh. Curr Sci 78:1105–1112

Nylander W (1860) Synopsis methodica lichenum I: 1850–1860 (Paris)

Nylander W (1867) Lichenes Kurziani Calcutta. Flora 50:3–9

Nylander W (1869) Lichenes Kurziani Bengalensis. Flora 52:69–73

Nylander W (1873) Lichenes Insularum Andaman, Bulletin de la Société linnéenne de Normandie 117:162–183

Oksanen I (2006) Ecological and biotechnological aspects of lichens. Appl Microbiol Biotechnol 73:723–734

Pant G, Upreti DK (1993) The lichen genus *Diploschistes* in India and Nepal. Lichenologist 25:33–50

Pant G, Upreti DK (1999) Lichen genus *Stereocaulon* in India and Nepal. In: Mukerji KG, Chamola BP, Upreti DK, Upadhyay RK (Eds) Biology of lichens. Aravali Books International, New Delhi, pp 249–281

Purvis W (2000) What is a lichen? In: Purvis W, Jacqui M (Eds) Lichen. [Life series], The natural history museum. London, pp 5–18

Quraishi AA (1928) Lichens of the Western Himalayas, Proceedings of 15th Indian Science. Congress, p 228

Rai AN, Bergmann B (2002) Cyanolichens. Biol Environ: Proc R Ir Academ 102B:19–22

Rai H (2012) Studies on diversity of terricolous lichens of Garhwal Himalaya with special reference to their role in soil stability. PhD Thesis. H.N.B Garhwal University. Srinagar (Garhwal), Uttarakhand, India

Rai H, Khare R, Gupta RK, Upreti DK (2011) Terricolous lichens as indicator of anthropogenic disturbances in a high altitude grassland in Garhwal (Western Himalaya), India. Bot Orient 8:16–23

Rai H, Khare R, Nayaka S, Upreti DK (2013a) The influence of water variables on the distribution of terricolous lichens in Garhwal Himalayas. In: Kumar P, Singh P, Srivastava RJ (Eds) Souvenir, water & biodiversity, 22 May 2013, International day for biological diversity, Uttar Pradesh State Biodiversity Board 7:75–83

Rai H, Nag P, Khare R, Upreti DK, Gupta RK (2013b) Functional role of terricolous lichens in carbon-nitrogen dynamics of alpine habitats in Himalayas. Souvenir, UGC sponsored national conference on resource management & its sustainable use, 22–23 March 2013, Pt. L.M.S. Govt. P.G. College, Rishikesh (Dehradun), Uttarakhand, pp 15

Rai H, Nag P, Upreti DK, Gupta RK (2010) Climate warming studies in alpine habitats of Indian Himalaya, using lichen based Passive Temperature-enhancing System. Nat Sci 8:104–106

Rai H, Upreti DK, Gupta RK (2012) Diversity and distribution of terricolous lichens as indicator of habitat heterogeneity and grazing induced trampling in a temperate-alpine shrub and meadow. Biodivers Conserv 21:97–113

Rawat S, Upreti DK, Singh RP (2009) Lichen flora of Mandal and adjoining localities towards Ukhimath in Chamoli district of Uttarakhand. J Phytol Res 22:47–52

Richardson DHS (2002) Reflections on lichenology: achievements over the last 40 years and challenges for the future. Canadian J Bot 80:101–113

Rikkinen J (2009) Relations between cyanobacterial symbionts in lichens and plants. Microbiol Monogr 8:265–270

Rout J, Kar A, Upreti DK (2004) Lichens of Sessa Orchid sanctuary, West Kameng, Arunachal Pradesh. Phytotaxonomy 4:38–40

Rundel PW (1998) Water relations. In: Galun M (ed) CRC handbook of lichenology, vol. 2. CRC Press, Boca Raton, pp 17–36

Saklani A, Upreti DK (1992) Folk uses of some lichens in Sikkim. J Ethnopharmacol 37:229–233

Satya, Upreti DK (2009) Correlation among carbon, Nitrogen, Sulphur and physiological parameters of *Rinodina sophodes* found at Kanpur city, India. J Hazard Mater 169:1088–1092

Seaward MRD (1988) Contribution of lichens to ecosystems. In: Galun M (ed) CRC Handbook of Lichenology, vol. 2. CRC Press, Boca Raton, pp 107–129

Shahi SK, Shukla AC, Uperti DK, Dikshit A (2000) Use of lichens as antifungal drugs against superficial fungal infections. J Med Aromat Plant Sci 22:169–172

Sheik MA, Upreti DK, Raina AK (2006) Lichen diversity in Jammu and Kashmir, India. Geophytology 36:69–85

Shukla AC, Pandey KP, Dutta S, Shahi SK, Upreti DK, Dixit A (2003) Antifungal activity of lichens *Ramalina* sp. and *Stereocaulon* sp. against some storage fungi. Univ. Allahabad Studies (New Millennium Series) 2:7–9

Shukla V, Upreti DK (2007a). Heavy metal accumulation in *Phaeophyscia hispidula* en route to Badrinath, Uttaranchal, India. Environ Monit Assess 131:365–369

Shukla V, Upreti DK (2007b) Physiological response of the lichen *Phaeophyscia hispidula* (Ach.) Essl. To the urban environment of Pauri and Srinagar (Garhwal), Himalayas, India. Environ Pollut 150:295–299

Shukla V, Upreti DK (2009) Polycyclic aromatic hydrocarbon (PAH) accumulation in lichen, *Phaeophyscia hispidula* of Dhra Dun city, Garhwal Himalayas. Environ Monit Assess 149:1–7

Shukla V, Upreti DK (2012) Air quality monitoring with lichens in India: heavy metals and polycyclic aromatic hydrocarbons. In: Lichtfouse E, Schwarzbauer J, Robert D (Eds) Environmental chemistry for a sustainable world. Springer Netherlands, pp 277–294

Singh KP (1977) Three new records of foliicolous lichens from India. Curr Sci 46:457–458.

Singh KP (1979) Lichen genus *Asterothyrium* Miill. Arg. in India. Curr Sci 48:267–268

Singh KP (1980) A New species of *Parmelina* from India. Bryologist 83:82–83

Singh KP (2011) Studies on Indian lichens during the last 50 years (1960–2010). Phytotaxonomy 11:120–140

Singh KP, Sinha GP (2010) Indian Lichens: an annotated checklist. Botanical Survey of India, Salt Lake City, Kolkata, p 507

Smith AL (1931) Lichens from northern India. Trans Br Mycol Soc 16:128–132

Subramanian SS, Ramakrishnan S (1964) Amino acids of *Peltigera canina*. Curr Sci 33:522

Taylor T (1852) New lichens principally from the herbarium of Sir Willam J. Hooker. Hooker's Lond J Bot 6:148–197

Tiwari P, Rai H, Upreti DK, Trivedi S, Shukla P (2011a) Antifungal activity of a common Himalayan foliose lichen *Parmotrema tinctorum* (Despr. ex Nyl.) Hale. Nat Sci 9:167–171

Tiwari P, Rai H, Upreti DK, Trivedi S, Shukla P (2011b) Assessment of antifungal activity of some Himalayan foliose lichens against plant pathogenic fungi. Am J Plant Sci 2:841–846

Tretiach M, Brown DH (1995) Morphological and physiological differences between epilithic and epiphytic populations of the lichen *Parmelia pastillifera*. Ann Bot 75:627–632

Tschermak-Woess E (1988) The algal partner. In: M. Galun (ed) CRC Handbook of Lichenology, vol. 1. CRC Press, Boca Raton, pp 39–92

Upreti DK (1985a) Lichen genus *Cladia* from India and Nepal. J Econ Taxon Bot 7:722–724

Upreti DK (1985b) Studies on the lichen genus *Baeomyces* from India. Geophytology 15:159–163

Upreti DK (1987) Keys to the species of the lichen genus *Cladonia* from India and Nepal. Feddes Repert 98:469–473

Upreti DK (1988) A new species of lichen genus *Phylliscum* from India. Curr Sci 57:906–907

Upreti DK (1990) Lichen genus *Pyrenula* in India: I *Pyrenula subducta* spore type. J Hattori Bot Lab 68:269–278

Upreti DK (1991a) Lichen genus *Pyrenula* from India: The species with spores of *Pyrenula brunnea* type. Bulletin de la Société Botanique de France. Lettr Bot138:241–247

Upreti DK (1991b) Lichen genus *Pyrenula* from india: IV. Pyrenula cayenensis spore type. Cryptogam, Bryol Lichénol 12:41–46

Upreti DK (1992) Lichen genus *Pyrenula* from India: VII. *Pyrenula mastophora* spore type. Feddes Repert 103:279–296

Upreti DK (1993a) Lichen genus *Pyrenula* from india: II. *Pyrenula camptospora* – spore type, III – Pyrenula pinguis -spore type. Acta Bot Gallica 140:519–523

Upreti DK (1993b) Notes on *Arthopyrenia* species from India. Bryologist 96:226–232

Upreti DK (1994) Notes on corticolous and saxicolous species of *Porina* from India with *Porina subhibernica* sp. nov. Bryologist 97:73–79

Upreti DK (1995) Loss of diversity in Indian lichen flora. Environmental conservation 22:361–363

Upreti DK (1996) Studies in Indian Lichenology-An overview. In: Jain SK, Kapoor SL, Rao RR (Eds) Ethnobiology in Human welfare. pp 413–414

Upreti DK (1997a) Diversity of Himalayan lichens. In: Sati SC, Saxena J, Dubey RC (Eds) Himalayan Microbial Diversity (Part 2). [Recent Researches in Ecology, Environment and Pollution] Today and Tomorrow's Printers & Publishers, New Delhi, pp 339–347

Upreti DK (1997b) Notes on corticolous K + yellow species of *Lecanora* in India. Feddes Repert 108:185–203

Upreti DK (1997c) Notes on Saxicolous species of the *Lecanora subfusca* group in India. Bryologist 101:256–262

Upreti DK (1998) A key to the lichen genus *Pyrenula* from India, with nomenclatural notes. Nova Hedwig 66:557–576

Upreti DK (2001a) Himalayan lichens and their exploitation. In: Pande PC, Samant SS (Eds) Plant diversity of the Himalaya. Gyanodaya Prakashan, Nainital, pp 95–100.

Upreti DK (2001b) Taxonomic, pollution monitoring and ethnolichenological studies on Indian lichens. Phytomorphology 51:477–497

Upreti DK (2001c) Utilization of lichens as human food and condiments. In: Pandey, G., Singh, J. (Eds) Natural resource management and conservation. pp 93–98

Upreti DK, Büdel B (1990) The lichen genera *Heppia* and *Peltula* in India. J Hattori Bot Lab 68:274–284

Upreti DK, Chatterjee S, Divakar PK (2004) Lichen flora of Gangotri and Gomukh areas of Uttaranachal, India. Geophytology 34:15–21

Upreti DK, Divakar PK (2010) A new species in the lichen genus *Sticta* (Schreb.) Ach. (Lobariaceae) from the Indian subcontinent. Nova Hedwig 90:251–255

Upreti DK, Divakar PK, Nayaka S (2005) Commercial and ethnic use of lichens in India. Econ Bot 59:269–273

Upreti DK, Joshi Y, Nayaka S, Joshi S (2010) New records of squamulose lichens from western Himalayas. Geophytology 38:85–91

Upreti DK, Nayaka S (2000) An enumeration of lichens from Himachal Pradesh. In: Chauhan DK (ed) Recent trends in biology, Prof. D.D. Nautiyal commemoration vol. Botany department, Allahabad University, Allahabad, India, pp 15–31

Upreti DK, Nayaka S (2008) Need for creation of lichen garden and sanctuaries in India. Curr Sci 94:976–978

Upreti DK, Nayaka S, Yadava V (2002) An enumeration and new records of lichens from Sirmaur district, Himachal Pradesh, India. Phytotaxonomy 2:49–63

Upreti DK, Negi HR (1995) Lichens of Nanda Devi Biosphere reserve, Uttar Pradesh, India. J Econ Taxon Bot 19:627–636

Upreti DK, Negi HR (1996) Folk use of *Thamnolia vermicularis* (Swartz.) Ach. in Schaerer in Lata village of Nanda Devi Biosphere Reserve. Ethnobotany 8:83–86.

Upreti DK, Negi HR (1998) Lichen flora of Chopta–Tungnath, Garhwal Himalayas, India. J Econ Taxon Bot 22:273–286

Upreti DK, Pandey V (1994) Heavy metals of Antarctica lichens-1. *Umbilicaria*. Feddes Repet 105:197–199

Upreti DK, Pandey V (1999) Determination of heavy metals in Lichens growing on different ecological habitats in Schirmacher Oasis, East Antarctica. Spectrosc Lett 33:435–444

Upreti DK, Pant V, Divakar PK (2001) Distribution of lichens in Askote Sandev botanical hotspot of Pithoragarh district, Uttar Pradesh. Phytotaxonomy 1:40–45.

Valarmathi R, Hariharan GN, Venkataraman G, Parida A (2009) Characterization of a non-reducing polyketide synthase gene from lichen *Dirinaria applanata*. Phytochemistry 70:721–729

Valarmathi R, Hariharan UN (2007) Soredial culture of *Dirinaria applanata* (Fee) Awasthi: observations on developmental stages and compound production. Symbiosis 43:137–142

Verma N, Behera BC, Joshi A (2012a) Studies on nutritional requirement for the culture of lichen *Ramalina nervulosa* and *Ramalina pacifica* to enhance the production of antioxidant metabolites. Folia Microbiol 57:107–114

Verma N, Behera BC, Makhija U (2008) Antioxidant and hepatoprotective activity of a lichen *Usnea ghattensis in vitro*. Appl Biochem Biotechnol 151:167–181

Verma N, Behera BC, Parizadeh H, Sharma B (2011a) Bactericidal activity of some lichen secondary compounds of *Cladonia ochrochlora, Parmotrema nilgherrensis* & *Parmotrema sanctiangelii.* Int J Drug Dev Res 3:222–232

Verma N, Behera BC, Sharma BO (2012b) Glucosidase inhibitory and radical scavenging properties of lichen metabolites salazinic acid, sekikaic acid and usnic acid. Hacet J Biol Chem 40:7–21

Verma N, Behera, BC, Makhija UV (2011b) Studies on cytochromes of lichenized fungi under optimized culture conditions. Mycoscience 52:65–68

Will-Wolf S, Esseen PA, Neitlich P (2002) Methods for monitoring biodiversity and ecosystem function. In: Nimis PL, Scheidegger C, Wolseley PA (Eds) Monitoring with lichens-monitoring lichens. [NATO Science Series IV: earth and environmental science vol. 7]. Kluwer Academic Publishers, Dordrecht, pp 147–162

Chapter 2
Distribution Ecology of Soil Crust Lichens in India: A Comparative Assessment with Global Patterns

Roger Rosentreter, Himanshu Rai and Dalip Kumar Upreti

1 Introduction

Lichens that occur either directly in soil, sand, peat/humus, or in habitats dominated by soil (e.g., on soil accumulated in rock crevices, on ground in mosses which in turn get rooted to the soil/sand or on degraded plant remains) constitute a unique habitat subset of a lichen community, known as terricolous lichens (Scheidegger and Clerc 2002). Terricolous lichens along with mosses and cyanobacteria form an intimated associative functional entity, often referred to as biological soil crust (BSC). Soil crusts and their component organisms are linked closely to enhanced soil and landscape stability in arid and semiarid areas. It is logical therefore, to view their presence as indicators of good landscape health. In India, both the arid desert in its northwestern part, the grasslands, and steppes from the foothills to the alpine regions in Himalayas contain habitats suitable for growth of terricolous lichens (Rai et al. 2011, 2012).

The Thar desert, in the western state of Rajasthan, is a climatically dry region and experiences frequent droughts. Comparatively the most densely populated desert in the world, this area holds a high livestock population (Sinha et al. 1996; Sharma and Mehra 2009). Therefore, this desert has a history of intense human pressure in the form of overgrazing by livestock, and fuel wood collecting (Sharma and Mehra 2009). Soils are generally sandy and sandy-loam in texture and high in soluble salts (Gupta 1968). The low nutrients in the soil, its high sandy texture, low humidity, and intense zooanthropogenic pressure, inhibit large scale growth of terricolous lichens in the region. The terricolous lichen growth in this desert region is restricted to some

R. Rosentreter (✉)
Idaho Bureau of Land Management, 2032 S. Crystal Way,
Boise, ID, 83706 USA
e-mail: Roger.rosentreter0@gmail.com

H. Rai · D. K. Upreti
Lichenology laboratory, Plant Diversity, Systematics and Herbarium Division
CSIR-National Botanical Research Institute,
Rana Pratap Marg, Lucknow
Uttar Pradesh-226001, INDIA

H. Rai, D. K. Upreti (eds.), *Terricolous Lichens in India,*   21
DOI 10.1007/978-1-4614-8736-4_2, © Springer Science+Business Media New York 2014

high altitude, moist habitats such as Mount Abu (1,220 m). The dominant soil crust species in this region are those of bipartite cyanolichen (having cyanobacteria as the only photosynthetic partner) genus *Collema* (*C. ryssoleum*, *C. texanum*, and *C. thamnodes*) along with sporadic occurrence of *Phaeophyscia hispidula*.

The temperate-alpine habitats (1,500 to >3,500 m) in the Himalayas, with steep, inclined mountainous terrains, contain flat alpine grasslands locally known as *Bug-yals*. These alpine grasslands harbour biological soil crusts, occasionally dominated by lichens (Rai et al. 2011, 2012). At lower altitude (1,800–2,000 m) lichen-domi-nated BSCs are rare but when they are present, they are not very diverse, whereas at mid- (≤3,400 m) to higher (>3,500 m) altitudes, lichen-dominated soil crusts are very diverse (Fig. 2.1). The temperate-alpine region of the Himalayas, are expected to be some of the most highly impacted lands by future climate changes as well as zooanthropogenic pressures (Rai et al. 2010, 2011, 2012). In *Bugyals*, sites with mosses tend to harbour a substantial population of lichens in the soil surface (Rai et al. 2011, 2012; Rai 2012). Lichens are abundant in the *Rhododendron*-rich mid-dle-montane altitudes (3,000–3,400 m). The most abundant lichen in the grasslands is a tripartite (a fungus, having two photosynthetic partners—a green algae and a cyanobacteria) fruticose cyanolichen *Stereocaulon foliolosum*, which has low palat-ability (Ahti 1959, 1964; Ahti et al. 1973) and is a well established nitrogen fixer (Fig. 2.1). The reason *Stereocaulon* is so common may be due to its low palatability, and resistance to grazing pressure (Rai et al. 2012), and its ability to reproduce by fragmentation. The second most abundant lichen is *Cladonia* spp. (*C. coccifera*, *C. pyxidata*)-a compound lichen growth form (squamules as primary thallus bearing erect fruticose body as secondary thallus), which along with *Stereocaulon* spp. are more adapted to harsh alpine climate (Sheard 1968; Rai et al. 2011, 2012; Fig. 2.1). The habitats of Himalayas, usually inhabited by terricolous lichens, are regions with harsh climate, characterized by regular orographic precipitation, longer periods of snow fall, higher UV radiation, and freezing minimum (−30 ^{0}C) temperatures (Rai et al. 2011, 2012; Khare et al. 2010). Middle elevation (300–3,400 m) sites ap-pear more favorable for soil lichen cover (Baniya et al. 2010), yet grazing pressure from livestock at middle altitudes and decrease in soil cover at higher elevations (>3,500 m) appears to limit their cover (Rai et al. 2011, 2012). However, from the foothills to the subalpine grasslands any lithic (rocky or shallow) soils have biologi-cal soil crusts. The species that occur in these sites are similar in composition to soil crusts around the world in similar sites. These lithic shallow soils are sometimes referred to as "bald" since they occur beyond the tree line and are dominated by grasses and herbaceous plants of *Asteraceae*.

2 Ecological Function of Soil Crusts and Terricolous Lichens

Biological soil crusts (BSCs) are a complex mosaic of cyanobacteria, green algae, lichens, mosses, microfungi, and other bacteria (Belnap et al. 2001). BSCs have a major influence on terrestrial ecosystems, including soil fertility and soil stability

Fig. 2.1 a, b Lichen-dominated soil crust at lower altitude (1,800–2,000 m), with lower species diversity, dominated by *Cladonia coniocraea* (**b**). **c–e** Lichen-dominated soil crusts at mid- (≤3,400 m) to higher (>3,500 m) altitudes, with higher species diversity, **d** *Cladonia pyxidata*, **e** *Stereocaulon foliolosum*

(Belnap 2003). In the arid and semiarid habitats of the world, they may constitute as much as 70 % of living cover (Belnap 1994). In the western USA, BSCs are critical components of healthy ecosystems (Rosentreter and Belnap 2003). Biological soil crusts cover the arid and semiarid deserts around the world and have been studied in the western portions of North America the most (Rosentreter and Belnap 2003). The benefits of BSCs in the landscape cannot be understated. These benefits include: soil building, erosion reduction, greater water capture and retention by soils, lessening of severity of dust storms, control of invasive plants through inhibition of germination, soil temperature amelioration, help in soil microbial growth, moderation of fire events through reduction of fine fuels, and improving perennial plant growth (Belnap et al. 2001; Sofronov et al. 2004; Fig. 2.2). Lichen-dominated BSCs, as

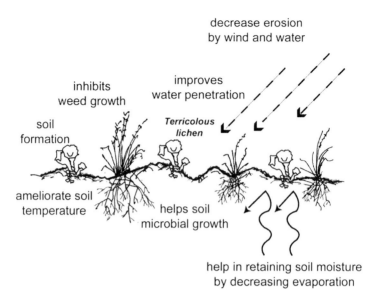

Fig. 2.2 Ecological benefits of lichen-dominated soil crusts in landscape ecology

a constituent of cryptogamic ground cover (CGC), along with cryptogamic plant cover (CPC) constitute a global continuum of cryptogamic cover, acting as major sink of atmospheric CO_2 and nitrogen accounting for about 7 % of net primary production and about half of biologically fixed nitrogen in terrestrial biomes (Elbert et al. 2009, 2012).

Biological soil crust performs a number of roles in semiarid ecosystems. Structurally, soil crusts bind soil particles. Functionally, crusts alter the soil chemistry and increase the rates of decomposition. As such, crusts are considered to be ecosystem engineers (Bowker et al. 2004).

Studies on the functional role of soil crust lichens (terricolous lichens) in Himalayan habitats have shown that terricolous lichen plays a key role in the maintenance of a number of physicochemical properties of soil (i.e., aggregate stability, soil temperature, soil microbial respiration, and soil carbon–nitrogen content; Rai 2012). Besides influencing the soil properties, terricolous lichen distribution in Himalayas is also intricately correlated with the soil's physicochemical properties and competition with other ground vegetation (Rai 2012).

3   Ecological Distribution Patterns of Soil Lichens

Some of the same characteristics that influence the distribution of vascular plant taxa also influence BSCs and terricolous lichen development and distribution (Kaltenecker 1997). Relative cover of BSCs in various generalized vegetation types varies according to various topographical and hydrological properties of habitats. Lichen-dominated BSCs tend to lack vegetation types that occur on seasonally flooded soils as flooded soils create anaerobic conditions which are not tolerated by lichens. Saline soils also lack lichen cover, although mosses are sometimes present if the salt concentration is not too high. Dense vegetation types lack significant biological soil crust cover as the closed canopies of vascular vegetation and accumulating plant litter create too much shade on the soil surface. Other vegetation types support higher biological soil crust cover unless their soil surfaces are highly disturbed or the current vegetation is in an early or recovering successional stage.

Biological soil crust lichens sometime display specific site affinities on both fine and gross scales. Some terricolous lichens have a high affinity for calcareous substrates (e.g., *Aspicilia hispida, Buellia elegans, Caloplaca tominii, Collema tenax, Psora decipiens, Toninia sedifolia*; McCune and Rosentreter 2007). These calcicolous lichens indicate free calcium carbonate in the soil, and are good indicators of soil pH (McCune and Rosentreter 2007).

Terricolous lichens form natural replacement series along the same elevation and moisture gradients that influence vascular plants. For example, at low elevations in India, the cyanolichen genus *Collema* fixes nitrogen while at mid-elevations the genus *Stereocaulon* is the nitrogen fixer. Functionally, the gelatinous (blue-green algae containing) terricolous lichens are all nitrogen fixers, are more resistant and resilient to disturbance, and are common in arid calcareous sites or mesic noncalcareous sites (McCune and Rosentreter 2007). Therefore the terricolous lichen community composition will differ between the hot, dry deserts with higher calcareous soil and cooler, moist foothills-alpine habitats with mostly acidic soils (Rai et al. 2012; Table 2.1). Sites with frigid soil temperature regimes (the mean annual temperature is < 8 C) lack significant cover of gelatinous lichens (Belnap et al. 2001). The genera *Peltigera* and *Massalongia* tend to be the common genera at these higher, more frigid elevations in North America while the genus *Stereocaulon* is common worldwide in frigid, high elevations (DeBolt and McCune 1993; Rai et al. 2012; Rai 2012). *Stereocaulon* is a common genus in open habitats at high elevations in India as well as Alaska, Canada and parts of South America (Rosentreter and Belnap 2003; Rai et al. 2011, 2012). Some species display a shift in substrate preference in different ecoregions. For example, in the Great Basin desert of North America, *Leptochidium albociliatum* occurs on mosses while in the moister Columbia Basin it is more common on bare mineral soil (Rosentreter 1986; Ponzetti et al. 1998).

4   Indicator Value of Terricolous Lichens

Terricolous lichens are good indicators of old-growth/late succession habitats (McCune and Rosentreter 2007). The dual gradient theory proposed by McCune (1993) for lichen species succession in forested habitats applies well to arid and semiarid regions where species respond to both time (age) and moisture in similar successional trajectories. Therefore, the length of time since the last major

Table 2.1 Comparison of the common lichens in the low elevation arid grassland/deserts versus the foothills-montane-alpine grasslands and steppes (Rai et al. 2012)

Arid low-elevation grasslands	Foothills-alpine grasslands and steppes
Collema tenax	*Cetrelia olivetorum*
Leptogium spp.	*Cladonia* spp.
Placidium spp.	*Everniastrum cirrhatum*
Endocarpon pusillium	*Lepraria* spp.
Lepraria spp.	*Stereocaulon foliolosum*
Heppia spp.	*Stereocaulon pomiferum*

disturbance of a site or an increase in effective soil moisture will both provide suitable ecological conditions to support specific lichen species. Late successional indicator species in arid steppe vegetation-type habitats include: *Acarospora schleicheri*, *Massalongia carnosa*, *Pannaria cyanolepra*, and *Trapeliopsis* species (Table 2.2). Some lichens only occur in stable, late successional communities because they only grow upon other lichen or moss species, e.g., *Acarospora schleicheri* spores germinate and grow upon the lichen, *Diploshistes muscorum*, which only grows upon the lichen genus *Cladonia*. Therefore, *Cladonia* can be considered a keystone species influencing the diversity of the site. *Massalongia carnosa* primarily grows on mosses and is not present until mosses become well distributed within a site (McCune and Rosentreter 2007). Another example from western North America is the rare lichen, *Texosporium sancti-jacobii* that is restricted to old-growth plant communities and occurs only on decaying organic matter (McCune and Rosentreter 1992). Other lichen species that commonly occur on decaying organic matter are: *Buellia papillata*, *B. punctata*, *Caloplaca* spp., *Lecanora* spp., *Megaspora verrucosa*, *Ochrolechia upsaliensis*, *Placynthiella* spp., and *Phaeophyscia decolor* (McCune and Rosentreter 2007).

A few common terricolous lichens that establish and grow quickly are early successional indicator species. The most common early successional lichen indicators are: *Collema tenax*, *Caloplaca tominii*, *Lepraria* spp., *Placidium* spp., and *Ochrolechia inaequatula*. Most of these species reproduce asexually, a life-history strategy which increases the probability of establishment (Rosentreter 1995). The presence, absence, and abundance of early- or late-successional species can provide information regarding the disturbance history of a site. Several early-successional species at a site indicate a recently disturbed site while several late-successional species at a site indicate less intensity or fewer disturbances in the recent history of the site. This information, combined with data on vascular plant community composition, can assist the land manager in understanding the disturbance history, potential productivity, and integrity of a site. Due to this ecological history at a site, the phase, "Lichens don't lie" is often used to demonstrate that the lack of late-successional lichens indicating that domestic grazing animals have disturbed the area.

The potential soil crusts and lichen cover for a given site are influenced by factors such as, associated vegetation, humidity/precipitation, soil texture, ecological successional stage, fire incidence frequency, and grazing pressure (Table 2.3). Soil

	Early successional soil lichens	Late successional soil lichens
Table 2.2 Early and late successional terricolous lichens in arid habitats	Collema tenax	Acarospora schleicheri
	Cladonia spp.	Massalongia carnosa
	Placidium spp.	Pannaria spp.
	Caloplaca toninii	Trapeliopsis spp.
	Ochrolecia inaequatula	Cetrelia spp.
	Lepraria spp.	Cetraria spp.

texture and vegetation of the site are critical factors (Kaltenecker et al. 1999). The factors listed are closely related and are components of the ecological site; however ,variation in any one factor can influence biological soil crusts crust cover and its relative importance to the ecological stability of the site (Hilty et al. 2004). In general, ecological sites dominated by bunch-grass grasslands will consistently have a well-developed "high" biological soil crust cover (Rosentreter and Belnap 2003). Soil texture of a site influences the stability of soil matrix, e.g., arid grasslands communities occurring on calcareous, gravelly loams and silt loams (such as alluvial deposits) have well developed lichen crusts that occupy fine-textured, mineral soil within a stabilized gravel matrix and are protected from livestock tramping (Table 2.3). The livestock impact on soil crust lichens is determined by the season of use and utilization intensity. Vegetation utilization is representative of animal stocking rates (number of sheep etc.) or length of grazing period. Severe to high utilization is indicative of localized concentration of animals and heavy trampling. Again, trampling impacts will be more severe if the soils are dry. Although the moisture required by soil crust community is lower than that needed by vascular plants, it is an important influencing factor in growth and development of diverse terricolous lichen community. Soil crusts are fragile when dry (dormant), but quite pliable when moist. Least impact occurs when the crust is moist or frozen but not saturated (Belnap et al. 2001).

5 Ecological Patterns of Terricolous Lichens in India

Terricolous lichens in India, more or less share the same distribution ecology patterns, which globally influence soil crusts. Soil crust formation in India is greatly influenced by grazing-induced disturbances (Rai et al. 2012). Terricolous lichens in the foothills, and montane grasslands are better developed than desert regions, and more diverse due to the limited season that livestock can access and trample these habitats (Rai et al. 2012; Sinha et al. 1996). Livestock-induced trampling has been a dominant factor influencing the distribution of soil lichens in India (Rai et al. 2011, 2012; Rai 2012). The sandy textured soils of deserts are unstable when dry and have

Table 2.3 Site potential for the occurrence of biological soil crusts, evaluation sheet

Potential for biological soil crusts crust development based on physical and biological factors (based on site potential relative to herbaceous cover)				
Crust cover	High >25% cover	Moderate >25–15%	Low >15–3%	Very low >3%
Dominant vegetation type	Arid grasslands	Foothills, grasslands	Foothills and Montane steppe	Open woodland
Herbaceous plant density	Low	Low-moderate	Moderate-high	High
Dominant herbaceous life form	Bunchgrass	Bunchgrass	Bunchgrass/rhizomatous plants	Rhizomatous plants
Annual precipitation	<300 mm	300–360 mm	>360–410 mm	>410 mm
Soil surface texture (* most critical factor)	Silts-fine silt loams clays (excluding shrink/swell clays)	Loamy	Sandy, shrink-swell clays	Coarse sand gravel or broken rock (>80% rock fragment) desert pavement
Surface rock Cover	>1% stable, embedded rocks	<1% stable, embedded rocks	Mostly unstable rocks, not embedded	Only unstable rocks
Historical fire return interval	>50 years (this would include most bunchgrass habitats.)	25–50 years	10–25 years	<20 years
Current ecological condition	Mid- to late seral or potential natural community	Early- to mid-seral	Disturbed to early-seral	Disturbed or with weedy plant cover
Likelihood for a change in management to negatively impact the biological soil crusts	High	Medium	Low	Very Low
Current livestock season of use	Warm dry periods	Cool dry periods	Cool moist periods	Cold moist periods
Vegetation utilization by grazing	Severe to high >50%	Moderate <50%	Light <35%	Slight <25%

Impacts of livestock grazing on biological soil crusts crust are judged to be (1) significant, (2) present but not significant or (3) not present. This judgment is based upon site specific data or completion of the above table. This table is not a substitute for field knowledge or monitoring data. The table may be used to describe the analysis of the impacts of livestock grazing on biological soil crusts cover. This table can be used in most arid habitats. Most data collected relative to this table were developed in the Great Basin desert of the western USA

been severely impacted by livestock trampling whereas, fine textured soil in the montane grasslands and steppe are more tolerant.

Though the major distribution of terricolous lichens and soil crust in India is restricted to foothills and the Himalayan montane grasslands (Rai et al. 2012, 2013; Rai 2012), the western dryer desert region holds a restricted distribution of terricolous lichen genera *Collema* (*C. ryssoleum*, *C. texanum*, and *C. thamnodes*) and *Phaeophyscia hispidula*. Both the lichen genera are calcophilic (most of the soils of western India have higher calcium carbonate content) and are found in Mount Abu region (Awasthi 2007), which is relatively more moist and free from livestock grazing. The Himalayan terricolous lichen community is dominated by species of *Stereocaulon* and *Cladonia* followed by Peltigera *praetextata, P. rufescence,* and *Xanthoparmelia terricola* (Rai et al. 2012; Rai 2012). Himalayan soil crust lichens are better adapted to acidic soils.

6 Conclusion

Although the distribution of soil crust lichens in India broadly follows the global ecological patterns, showing striking taxonomic and ecological similarity with soil crust communities worldwide, their growth and development is constrained by constantly increasing grazing pressures. Furthermore, the rapidly changing land-use patterns in lichen-rich Himalayan habitats, is also proving detrimental to the survival of these soil crust species.

References

Ahti T (1959) Studies on the caribou lichen stands of Newfoundland. Ann Bot Soc Zool Bot Fenn 'Vanamo' 30:1–43

Ahti T (1964) Macrolichens and their zonal distribution in boreal and arctic Ontario, Canada. Ann Bot Fenn 1:1–35

Ahti T, Scotter GW, Vänskä H (1973) Lichens of the reindeer Preserve, Northwest Territories, Canada. Bryologist 76:48–76

Awasthi DD (2007) A compendium of the macrolichens from India, Nepal and Sri Lanka. Bishen Singh Mahendra Pal Singh, Dehra Dun, pp 580

Baniya CB, Solhøy T, Gauslaa Y, Palmer MW (2010) The elevation gradient of lichen species richness in Nepal. Lichenologist 42:83–96

Belnap J (1994) Cryptobiotic soil crusts: basis for arid land restoration (Utah). Restor Manage Notes 12:85–86

Belnap J (2003) Biological soil crusts and wind erosion. In: Belnap J, Lange OL (eds) Biological soil crusts: structure, function, and management, ecological studies. Series 150. Berlin, Springer, pp 339–347

Belnap J, Kaltenecker JH, Rosentreter R, Williams J, Leonard S, Eldridge D (2001) Biological soil crusts: ecology and management. USDI Bureau of Land Management National Science and Technology Center, Tech. Ref. 1730–2

Bowke MA, Belnap J, Rosentreter R, Graham B (2004) Wildfire-resistant biological soil crusts and fire-induced loss of stability in Palouse prairies, USA. App Soil Ecol 26:41–52

DeBolt A, McCune B (1993) Lichens of Glacier National Park, Montana. Bryologist 96:192–204

Elbert W, Weber B, Büdel B, Andreae MO, Pöschl U (2009) Microbiotic crusts on soil, rock and plants: neglected major players in the global cycles of carbon and nitrogen. Biogeosciences Discuss 6:6983–7015

Elbert W, Weber B, Burrows S, Steinkamp J, Büdel B, Andreae MO, Pöschl U (2012) Contribution of cryptogamic covers to the global cycles of carbon and nitrogen. Nat Geosci 5:459–462

Gupta RS (1968) Investigation on the desert soils of Rajasthan fertility and mineralo-gical studies. J Indian Soc Soil Sci 6:115–121

Hilty JH, Eldridge DJ, Rosentreter R, Wicklow-Howard MC, Pellant M (2004) Recovery of biological soil crusts following wildfire on the Snake River Plain, Idaho, U.S.A. J Range Manag 57:89–96

Kaltenecker J (1997) The recovery of microbiotic crusts following post-fire rehabilitation on rangelands of the western snake river plain. M.S. Thesis, Boise State University, Boise, ID, U.S.A.

Kaltenecker JK, Wicklow-Howard MC, Rosentreter R (1999) Biological soil crusts in three sagebrush communities recovering from a century of livestock trampling. In: McArthur ED, Ostler KW, Wambolt CL (comps.) Proceedings; 1998 August 12–14; Ephraim, UT. Ogden, UT: U.S. Department of Agriculture, Forest Service, Rocky Mountain Research Station: 222–226. Shrubland ecotones (1999)

Khare R, Rai H, Upreti DK, Gupta RK (2010) Soil Lichens as indicator of trampling in high altitude grassland of Garhwal, Western Himalaya, India. Fourth National Conference on Plants & Environmental Pollution, 8–11 Dec. 2010, pp 135–136

McCune B (1993) Gradients in epiphyte biomass in three *Pseudotsuga-Tsuga* forests of different ages in western Oregon and Washington. Bryologist 96:405–411

McCune B, Rosentreter R (1992) *Texosporium sancti-jacobi*, a rare western North American lichen. Bryologist 95:329–333

McCune B, Rosentreter R (2007) Biotic soil crust lichens of the Columbia Basin. Monogr N Am Lichenol 1:1–105

Ponzetti J, Youtie B, Salzer D, Kimes T (1998) The effects of fire and herbicide on microbiotic crust dynamics in high desert ecosystems. Corvallis: U.S. 107 Department of Agriculture, Forest and Rangeland Ecosystem Science Center, Oregon State University

Rai H (2012) Studies on diversity of terricolous lichens of Garhwal Himalaya with special reference to their role in soil stability. PhD Thesis. H.N.B Garhwal University. Srinagar (Garhwal), Uttarakhand, India

Rai H, Khare R, Gupta RK, Upreti DK (2011) Terricolous lichens as indicator of anthropogenic disturbances in a high altitude grassland in Garhwal (Western Himalaya), India. Bot Orientalis 8:16–23

Rai H, Nag P, Upreti DK, Gupta RK (2010) Climate warming studies in alpine habitats of Indian Himalaya, using lichen based passive temperature-enhancing system. Nat Sci 8:104–106

Rai H, Upreti DK, Gupta RK (2012) Diversity and distribution of terricolous lichens as indicator of habitat heterogeneity and grazing induced trampling in a temperate-alpine shrub and meadow. Biodivers Conserv 21:97–113

Rai H, Khare R, Nayaka S, Upreti DK (2013) The influence of water variables on the distribution of terricolous lichens in Garhwal Himalayas. In: Kumar P, Singh P, Srivastava RJ (eds) Souvenir, water & biodiversity, vol 7, pp 75–83. International day for biological diversity, Uttar Pradesh State Biodiversity Board, 22 May 2013

Rosentreter R (1986) Compositional patterns within a rabbitbrush (*Chrysothamnus*) community of the Idaho Snake River Plain. in McArthur EE Welch BL, comps (eds) Proceedings of the Symposium: Biology of *Artemisia* and *Chrysothamnus*. Gen. Tech. Report INT-200, Ogden, UT

Rosentreter R (1995) Lichen diversity in managed forests of the Pacific Northwest, USA. 103–124 pp. In: Conservation of Lichenized Fungi. Scheidegger C, Wolseley PA, Thor G (eds) 1995. Mittenelumgen der Eidgenossischen Forschungsanstalt fur Wald. Schenee und Landschaft 70, 1:1–173. Birmensdorf, Switzerland

Rosentreter R, Belnap J (2003) Biological soil crusts of North America. In Biological Soil Crusts: Structure, Function, and Management, Ecological Studies Series 150. In: Belnap J, Lange OL (eds) Berlin, Springer, pp 31–50

Scheidegger C, Clerc P (2002) Erdbewohnende Flechten der Schweiz. In Rote Liste der gefährdeten Arten der Schweiz: Baum- und erdbewohnende Flechten, pp 75–108

Sharma KK, Mehra SP (2009) The Thar of Rajasthan (India): ecology and conservation of a desert ecosystem. In: Sivaperuman C et al. (eds) Faunal ecology and conservation of the Great Indian Desert. Springer, Berlin

Sheard JW (1968) Vegetation pattern on a moss-lichen heath associated with primary topographic features on Jan Mayen. Bryologist 71:21–29

Sinha KR, Bhatia S, Vishnoi R (1996) Desertification control and rangeland management in the Thar desert of India. RALA Report No. 200, 115–123

Sofronov MA, Volokitina AV, Kajimoto T, Uemura S (2004) The ecological role of moss-lichen cover and thermal amelioration of larch forest ecosystems in the northern part of Siberia. Eurasian J For Res 7:11–19

Chapter 3
Terricolous Lichens in Himalayas: Patterns of Species Richness Along Elevation Gradient

Chitra Bahadur Baniya, Himanshu Rai and Dalip Kumar Upreti

1 Introduction

Lichens are the primary colonizers of the home planet along with non-lichenized fungi, algae, bacteria, cyanobacteria, and mosses. Diverse morphological growth forms and tolerant physiology provide lichens an edge over other cryptogams, enabling them to colonize nearly all of the terrestrial habitats of our planet. Terricolous lichens (highly sensitive) are important subsets of lichen community, as they highly sensitive to climatic, topographic, and anthropogenic factors (Will-Wolf et al. 2002). Terricolous lichens along with cyanobacteria, algae, microfungi, and bryophytes (in different proportions) sometimes constitute an intimate functional entity referred to as biological soil crusts (BSCs) (Belnap et al. 2001a). The biological soil crust is the heterogeneous material that forms after colonization, nutrient enrichment, and stabilization of the soil lichens (Eldridge 1996). BSCs impart spatial and temporal patterns at different scales determining biological diversity that begins from harsh to congenial environments of any stages of succession. Terricolous lichen diversity in BSCs is closely associated with habitat characteristics, and therefore indicates the health of ecosystem (Pellant et al. 2001).

C. B. Baniya (✉)
Central Department of Botany, Tribhuvan University,
Box 15142 KPC 785, Kirtipur, Kathmandu, Nepal
e-mail: cbbaniya@gmail.com; cb.baniya@cdbtu.edu.np

H. Rai · D. K. Upreti
Lichenology laboratory, Plant Diversity, Systematics and Herbarium Division
CSIR-National Botanical Research Institute,
Rana Pratap Marg, Lucknow
Uttar Pradesh-226001, INDIA
e-mail: himanshurai08@yahoo.com

H. Rai, D. K. Upreti (eds.), *Terricolous Lichens in India,*   33
DOI 10.1007/978-1-4614-8736-4_3, © Springer Science+Business Media New York 2014

2   Terricolous Lichens: Role in Ecosystem Services

Terricolous lichens by virtue of their requirement of greater environmental stability
are highly sensitive indicators of the overall ecosystem functioning and various environmental disturbances (Eldridge and Tozer 1997; Grabherr 1982; Scutari et al.
2004; Lalley and Viles 2005; Motiejûnaitë and Wiesùaw 2005; Lalley et al. 2006a; Clair et al.
2007; Rai et al. 2011, 2012). The ecological and geomorphic importance of terricolous
lichen communities are related to the soil crust they inhabit. Terricolous lichens are
known to be sources of nitrogen and carbon fixation (Beymer Klopatek 1992; Evans
and Belnap 1999; Harper and Belnap 2001), vital soil stabilizers (Belnap and Gillette
1998; Eldridge 1998; Eldridge and Leys 2003), and providers of habitat and food
sources for other organisms (Zaady and Bouskila 2002; Lalley et al. 2006b), and are
known to influence growth of associated cryptogamic soil vegetation (Lawrey 1977;
Gardner and Mueller 1981; Sedia and Ehrenfeld 2005; Escurado et al. 2007; Lawrey
2009; Favero-Longo and Piervittori 2010). Terricolous lichen-dominated BSCs are an
indicator of physical, physiological, and chemical impacts after grazing, trampling,
climate change, and pollution (Belnap and Eldridge 2001; Belnap et al. 2001a, b).

3   Terricolous Lichens in the Himalayas

Himalayan habitats show diversity both in climate and habitat conditions, where
they are more moist in the eastern Himalayan alpine shrubs and meadows, which
act as a transit into drier central Tibetan plateau alpine steppe of central Tibet in
the north. The stressed climate (i.e., higher environmental lapse rate, high wind
velocity, high UV radiation, low atmospheric pressure, and low precipitation), and
the delimiting nutrient and exposure regime of Himalayas, support relatively simple
ecosystems, characterized by limited trophic levels and relatively, very few plant
growth forms and species (Rai et al. 2010). Despite these constrains, alpine habitats
of the Himalayas harbor some of the unique biodiversities of the region, which are
vital for the overall ecosystem functioning and stability. Himalayan habitats are rich
in lichen biodiversity, which constitute a substantial amount of cryptogamic ground
vegetation (Upreti 1998). Major distribution of Terricolous lichens in the Himalayas
ranges from temperate (1,500–3,000 m) to alpine (>3,000 m) habitats (Baniya et al.
2010; Rai et al. 2011). Altitudinal distribution patterns of terricolous lichens of the
Himalayan region are addressed in few region-specific lichenological investigations (Pinokiyo et al. 2008; Baniya et al. 2010; Huang 2010; Rai et al. 2012).

4   Objectives

The present work has been undertaken to elucidate an elevational richness pattern of
the Himalayan terricolous lichens with reference to (i) mid-dominance peak of terricolous lichen diversity along the gradient; (ii) dominant terricolous lichen families

that follow a similar richness pattern, and (iii) probable determining factors to explain each of these patterns.

5   Materials and Methods

5.1   Study Area

The Himalayas are the youngest fragile mountains of Cenozoic origin. Present vegetation is believed to be the Pleistocene formation established after Miocene orogeny
(Singh and Singh 1987). The Himalayas arch the massive Indian subcontinent including Nepal, India, Burma, Bhutan, and Sri Lanka, which leaves the Tibetan plateau
on the backside. The Himalayan range is about 2,500 km long and passes all the way
from Burma in the east to Afghanistan in the west. The Himalayas lie between 84°E
and 84°W longitude (Singh and Singh 1987). It is the confluent of Austro-Poly-
nesian, Malayo-Burman, Sino-Tibetan, Euro-Mediterranean, and African elements.
These topographies with environment are believed to be suitable to all these flora
and fauna since ancient history. Soil lichens collected, described, deposited, and published mainly from the Himalayan region of India and Nepal are considered here.

The Himalayas comprise the longest and a continuous bioclimatic region of the
world. The elevation of the landscape varies from the sea level to the world's highest
peak, i.e., the Mount Everest, 8,848 m above sea level, within a short horizontal distan-
ce. Diverse topographies also harbor diverse vegetation typical of hot, wet tropics to ni-
val zones of the world. Hot and humid climates are characteristic features of landscapes
toward lower altitudes that slowly decrease with increasing altitude. Thus, temperature
and precipitation are two major gradients. Similarly, intensity of precipitation varies
differently along the Himalayas. Monsoon is the main source of rainfall that begins
from the Bay of Bengal and ends in the western Himalayas. The eastern Himalayas,
nearer to the Bay of Bengal, receive more rainfall during summer (May to September).
The intensity slowly decreases toward the western Himalayas. Terricolous lichens from
the Himalayas have witnessed and are indicators of these changes since prehistory.

5.2   Data Source

The main data source for this study is *A Compendium of the Macrolichens from
India, Nepal and Sri Lanka* (Awasthi 2007), the checklist of Nepalese lichens (Sharma 1995; Baniya et al. 2010). Along with the above-mentioned sources, data of
Cladonia of India and Nepal were supplemented by analysing about 700 specimens
lodged in lichen herbarium of the Council of Scientific and Industrial Research
(CSIR)-National Botanical Research Institute (LWG). The distribution pattern of
soil lichens of Nepal was crosschecked using data from Baniya et al. (2010).

The altitudinal ranges of the Himalayan soil lichens (100–6,000 m) were interpo-
lated after dividing them into 60 bands of 100 m each, and a complete data matrix

for all soil lichen species was assembled. Presence of a species indicates that the species occurs in, or has been collected in the past from, that altitudinal band and the absence means either that species does not occur or has previously not been collected from that altitude (Baniya et al. 2010).

A species is assumed to be present at all possible 100-m bands between its upper and lower altitudinal limits as recorded in the dataset. For example, a terricolous lichen *Baeomyces pachypus* that has elevational occurrences between 1800 to 3600 m in the literature falls between the 1,800- and 3,600-m bands throughout the subcontinent (Baniya et al. 2010). Endemic soil lichen species are encountered at a specific landscape or in a country that counts for all the Himalayas. The lichen species listed in the compendium or literature without any altitude reference were discarded. Such soil lichen species were around 30 in number.

Thus, the species richness that applies here is an estimate of the total number of soil lichen species and/or their family occurring at each 100-m altitudinal band throughout the Himalayas. This corresponds to macroscale study, which is closer to gamma diversity as mentioned by Whittaker (1972).

5.3 Data Analysis

Considering total lichen species richness, patterns related to, six dominant families as response variables and their elevations as a predictor variable were extracted by applying a cubic smooth spline (*s*) with the framework of generalized additive models (GAM) (Hastie and Tibshirani 1990; Heegaard 2004; Baniya et al. 2010). GAM, which is one of the most conservative, local regression methods has been used in this study, without priori. Response variables variables are counts; thus, the variance changes with the mean. Over dispersion in the dataset was corrected through an application of quasi-Poisson family of error, which has a logarithmic link function (Crawley 2006). Normal distribution in the error was tested after the basic Q–Q (quantile–quantity) diagnostic plots against residuals. The change in deviance follows the *F*-distribution. R 2.13.1 was used to analyse the data and smoothers (R Development Core Team 2011). The models were fitted with the library *GAM* (Hastie and Tibshirani 1990).

6 Results

6.1 General Pattern of Species Richness

A total of 212[1] terricolous/soil lichen species, 1.06% of the total lichens reported from India alone, were found recorded from the Himalayas with altitudinal

---

[1] The data takes into account representative terricolous specimens lodged in CSIR-NBRI, lichenological herbarium-LWG.

distribution range. These terricolous lichens species belonged to 24 families and 54 genera. Their altitudes ranged from 100 to 6,000 m above sea level. *Cladoniaceae* was the dominant family with 53 species, followed by *Parmeliaceae* with 49 species (Table 3.1).

Total soil lichen species richness showed statistically significant unimodal relation with altitude. The highest number of soil lichen species (89) occurred at 2,400 m (Fig. 3.1; Table 3.2). The soil lichen species richness was found increasing from sea level (100 m) to 2,400 m and declining afterward.

6.2 Mid-Richness Peak of Dominant Families

The most dominant soil lichen families also showed unimodal responses to elevation at different altitudes of their highest richness (Fig. 3.2; Table 3.2). For instance, *Cladoniaceae* with 53 species showed unimodal relation of the highest richness (38 species) at 2,700–2,800 m (Fig. 3.2a; Table 3.2). *Parmeliaceae* with 49 species showed unimodal response with the highest richness (17 species) at 2,100 m (Fig. 3.2b; Table 3.2). *Peltigeraceae* (18 species) humped at 2,800 m by 9 species (Fig. 3.2c; Table 3.2), *Physciaceae* (17 species) humped at 1,700–1,800 m by 10 species (Fig. 3.2d; Table 3.2), *Collemataceae* (12 species) humped at 1,200–1,300 m by 6 species (Fig. 3.2e; Table 3.2), and *Stereocaulaceae* (12 species) humped at 3,500–3,600 m by 7 species (Fig. 3.2f; Table 3.2).

The study has found that 2,400 m is the most preferential altitude for soil lichens among 60 altitudinal bands. At this midaltitude (2,400 m) soil lichen species may be supported by all habitats, microenvironments, edaphic factors, and biogeography as well and vice versa.

7 Discussion

A quantitative analysis of the terricolous lichen community from the Himalayas has revealed that they are significantly greater in number and wider in distribution. These terricolous lichen communities are found distributed at all altitudes from 100 to 6,000 m. Moreover, 2,400 m altitude was the most preferable altitude and has abundant number of terricolous lichen species (Fig. 3.1). Not all lichen families behave the same in terms of appearance of maximum number of species with altitude. Terricolous lichens followed the general trend of elevational declining unimodal species richness pattern at larger spatial scales but differed with elevation of maximum richness. The fundamental reasons explaining for this pattern may share with the general pattern of other species richness but the specific cause would be specific to the terricolous lichen only.

Table 3.1 List of terricolous lichen species found in India and Nepal and their range of altitudinal distribution rank of genera and species

S. No. (Species)	S. No. (Genera)	Terricolous lichen species	S. No. (Family)	Family	No. of genera/ family	No. of spp./ family	Altitudinal range (m)	Frequency
1	1	Cladia aggregata	1	Cladoniaceae	2	53	1,500–3,600	22
2	2	Cladonia acuminata		Cladoniaceae			2,700–3,750	12
3		C. amaurocraea		Cladoniaceae			4,000–4,221	3
4		C. awasthiana		Cladoniaceae			1,500–3,500	21
5		C. borealis		Cladoniaceae			4,000–4,050	2
6		C. cariosa		Cladoniaceae			1,603–4,100	26
7		C. cartilaginea		Cladoniaceae			335–3,900	37
8		C. ceratophyllina		Cladoniaceae			2,286–3,810	16
9		C. cervicornis		Cladoniaceae			1,829–2,134	4
10		C. cfr fenestralis		Cladoniaceae			1,800	1
11		C. cfr ochrochlora		Cladoniaceae			2,100	1
12		C. chlorophaea		Cladoniaceae			1,500–4,425	30
13		C. coccifera		Cladoniaceae			2,010–4,420	25
14		C. coniocraea		Cladoniaceae			1,250–3,600	24
15		C. corniculata		Cladoniaceae			950–4,600	37
16		C. corymbescens		Cladoniaceae			1,100–4,420	34
17		C. crispata var. cetrariiformis		Cladoniaceae			3,250	1
18		C. delavayi		Cladoniaceae			1,618–4,250	28
19		C. didyma		Cladoniaceae			1,350–3,300	20
20		C. farinacea		Cladoniaceae			2,000–2,286	4
21		C. fenestralis		Cladoniaceae			50–4,724	47
22		C. fimbriata		Cladoniaceae			1,585–4,500	30
23		C. fruticulosa		Cladoniaceae			795–3,962	33
24		C. furcata		Cladoniaceae			560–4,250	38

Table 3.1 (continued)

S. No. (Species)	S. No. (Genera)	Terricolous lichen species	S. No. (Family)	Family	No. of genera/ family	No. of spp./ family	Altitudinal range (m)	Frequency
25		C. humilis		Cladoniaceae			2,250–2,850	7
26		C. indica		Cladoniaceae			350–700	4
27		C. laii		Cladoniaceae			2,700–3,900	13
28		C. luteoalba		Cladoniaceae			2,800–3,658	10
29		C. macilenta		Cladoniaceae			1,200–3,962	29
30		C. macroceras		Cladoniaceae			4,200	1
31		C. macroptera		Cladoniaceae			50–4,176	42
32		C. mauritiana		Cladoniaceae			609–2,750	23
33		C. mongolica		Cladoniaceae			1,615–3,750	23
34		C. nitida		Cladoniaceae			4,176	1
35		C. ochrochlora		Cladoniaceae			7,00–4,000	34
36		C. pocillum		Cladoniaceae			1,500–4,700	33
37		C. praetermissa		Cladoniaceae			1,505–1,550	2
38		C. pyxidata		Cladoniaceae			1,200–4,600	35
39		C. ramulosa		Cladoniaceae			1,050–4,000	30
40		C. rangiferina		Cladoniaceae			1,600–4,481	30
41		C. rei		Cladoniaceae			2,200–3,200	11
42		C. scabriuscula		Cladoniaceae			3,34–3,962	38
43		C. sinensis		Cladoniaceae			2,200	1
44		C. singhii		Cladoniaceae			1,200–3,048	20
45		C. squamosa		Cladoniaceae			1,000–4,221	33
46		C. stricta		Cladoniaceae			3,300–4,420	12
47		C. subconistea		Cladoniaceae			795–1,875	12
48		C. subradiata		Cladoniaceae			1,379–3,800	25

Table 3.1 (continued)

S. No. (Species)	S. No. (Genera)	Terricolous lichen species	S. No. (Family)	Family	No. of genera/family	No. of spp./family	Altitudinal range (m)	Frequency
49		*C. subsquamosa*		*Cladoniaceae*			1,295–2,567	14
50		*C. subulata*		*Cladoniaceae*			2,100–3,000	10
51		*C. turgida*		*Cladoniaceae*			2,210	1
52		*C. verticillata*		*Cladoniaceae*			1,000–4,100	32
53		*C. yunnana*		*Cladoniaceae*			1,800–3,963	23
54	3	*Allocetraria ambigua*	2	*Parmeliaceae*	18	49	4,875	1
55		*A. flavonigrescens*		*Parmeliaceae*			4,600–4,800	3
56		*A. sinensis*		*Parmeliaceae*			4,830	1
57		*A. stracheyi*		*Parmeliaceae*			3,000–5,000	21
58	4	*Bryoria bicolor*		*Parmeliaceae*			1,500–3,600	22
59		*B. implexa*		*Parmeliaceae*			3,700–4,000	4
60		*B. smithii*		*Parmeliaceae*			2,400–4,500	22
61		*B. tenuis*		*Parmeliaceae*			2,250–3,650	15
62	5	*Bulbothrix isidiza*		*Parmeliaceae*			800–1,500	8
63		*B. meizospora*		*Parmeliaceae*			1,500–2,250	9
64	6	*Canomaculina subtinctoria*		*Parmeliaceae*			450–2,500	21
65	7	*Cetraria aculeata*		*Parmeliaceae*			4,000	1
66		*C. islandica*		*Parmeliaceae*			3,450–4,500	11
67		*C. muricata*		*Parmeliaceae*			4,000	1
68		*C. nepalensis*		*Parmeliaceae*			4,500	1
69		*C. nepalensis*		*Parmeliaceae*			4,500	1
70		*C. nigricans*		*Parmeliaceae*			3,800	1
71	8	*Evernia mesomorpha*		*Parmeliaceae*			2,500–3,600	12
72	9	*Everniastrum cirrhatum*		*Parmeliaceae*			1,400–3600	23

Table 3.1 (continued)

S. No. (Species)	S. No. (Genera)	Terricolous lichen species	S. No. (Family)	Family	No. of genera/family	No. of spp./family	Altitudinal range (m)	Frequency
73		E. nepalense		Parmeliaceae			1,300–4,200	30
74		E. vexans		Parmeliaceae			1,200–2,250	12
75	10	Flavocetraria cucullata		Parmeliaceae			3,600–4,500	10
76		F. nivalis		Parmeliaceae			3,800–4,500	8
77	11	Flavocetrariella leucostigma		Parmeliaceae			3,900–4,500	7
78		F. melaloma		Parmeliaceae			3,900–4,200	4
79	12	Flavoparmelia caperata		Parmeliaceae			1,600–3,500	20
80	13	Hypogymnia delavayi		Parmeliaceae			3,400–5,600	23
81		H. fragillima		Parmeliaceae			3,400–3,780	5
82		H. hypotrypa		Parmeliaceae			3,600–4,050	6
83		H. physodes		Parmeliaceae			2,200–2,700	6
84	14	Hypotrachyna exsecta		Parmeliaceae			1,800–2,300	4
85		H. koyaensis		Parmeliaceae			1,500–2,100	7
86	15	Lethariella cladonioides		Parmeliaceae			3,600–5,400	19
87		L. cladonioides		Parmeliaceae			4,724	1
88	16	Melanelia hepatizon		Parmeliaceae			3,600	1
89		M. stygia		Parmeliaceae			3,300–3,600	4
90	17	Melanelixia fuliginosa		Parmeliaceae			1,500–2,500	11
91		M. villosella		Parmeliaceae			2,400–3,300	10
92	18	Parmelinella wallichiana		Parmeliaceae			500–3,000	26
93	19	Parmotrema crinitum		Parmeliaceae			1,300–2,250	11
94		P. grayanum		Parmeliaceae			900–2,400	16
95		P. mellissii		Parmeliaceae			850–2,200	14

Table 3.1 (continued)

S. No. (Species)	S. No. (Genera)	Terricolous lichen species	S. No. (Family)	Family	No. of genera/family	No. of spp./family	Altitudinal range (m)	Frequency
96		P. nilgherrense		Parmeliaceae			1,500–4,200	28
97		P. pseudocrinitum		Parmeliaceae			950–2,100	12
98		P. pseudonilgherrense		Parmeliaceae			1,600–3,500	20
99		P. reticulatum		Parmeliaceae			1,500–2,500	11
100		P. sancti-angelii		Parmeliaceae			1,410–2,100	8
101	20	Xanthoparmelia bellatula		Parmeliaceae			3,150–3,250	2
102		X. terricola		Parmeliaceae			3,900–4,500	7
103	21	Peltigera canina	3	Peltigeraceae	2	18	1,800–3,500	18
104		P. collina		Peltigeraceae			1,350–1,880	6
105		P. didactyla		Peltigeraceae			2,000–2,200	3
106		P. dolichorrhiza		Peltigeraceae			1,600–4,000	25
107		P. dolichospora		Peltigeraceae			3,000–4,100	12
108		P. elisabethae		Peltigeraceae			1,350	1
109		P. horizontalis		Peltigeraceae			2,400–3,450	12
110		P. leucophlebia		Peltigeraceae			3,800	1
111		P. malacea		Peltigeraceae			2,100–4,200	22
112		P. membranacea		Peltigeraceae			1,900–2,400	6
113		P. polydactylon		Peltigeraceae			1,950–2,920	10
114		P. polydactylon var. polydatylon		Peltigeraceae			1,950–3,480	16
115		P. praetextata		Peltigeraceae			1,950–3,600	17
116		P. pruinosa		Peltigeraceae			1,800	1
117		P. rufescens		Peltigeraceae			3,000–3,600	7
118		P. venosa		Peltigeraceae			2,100	1
119	22	Solorina bispora		Peltigeraceae			3,600–4,200	7

Table 3.1 (continued)

S. No. (Species)	S. No. (Genera)	Terricolous lichen species	S. No. (Family)	Family	No. of genera/family	No. of spp./family	Altitudinal range (m)	Frequency
120		S. simensis		Peltigeraceae			1,800–3,300	16
121	23	Anaptychia pseudoroemeri	4	Physciaceae	7	17	3,780	1
122	24	Awasthia melanotricha		Physciaceae			4,350–4,510	2
123	25	Diploicia canescens		Physciaceae			4,650	1
124	26	Heterodermia boryi		Physciaceae			1,500–3,000	16
125		H. diademata		Physciaceae			1,200–3,000	19
126		H. firmula		Physciaceae			1,200–2,200	11
127		H. hypocaesia		Physciaceae			800–2,500	18
128		H. japonica		Physciaceae			800–1,500	8
129		H. leucomelos		Physciaceae			800–1,500	8
130		H. obscurata		Physciaceae			800–2,500	18
131		H. pseudospeciosa		Physciaceae			800–2,500	18
132		H. tremulans		Physciaceae			800–1,500	8
133	27	Phaeophyscia ciliata		Physciaceae			1,800–3,500	18
134		P. constipata		Physciaceae			3,500	1
135	28	Physcia adscendens		Physciaceae			1,500–3,600	22
136		P. tribacoides		Physciaceae			800–1,500	8
137	29	Physconia muscigena		Physciaceae			1,500–2,500	11
138	30	Collema coccophorum	5	Collemataceae	2	12	3,800	1
139		C. poeltii		Collemataceae			3,900–4,000	2
140		C. rugosum		Collemataceae			800–2,500	18
141		C. subflaccidum		Collemataceae			1,500–3,000	16
142	31	Leptogium arisanense		Collemataceae			2,830	1
143		L. askotense		Collemataceae			2,330	1

Table 3.1 (continued)

S. No. (Species)	S. No. (Genera)	Terricolous lichen species	S. No. (Family)	Family	No. of genera/family	No. of spp./family	Altitudinal range (m)	Frequency
144		*L. corniculatum*		*Collemataceae*			800–1,500	8
145		*L. cyanescens*		*Collemataceae*			800–1,500	8
146		*L. denticulatum*		*Collemataceae*			800–1,500	8
147		*L. moluccanum*		*Collemataceae*			100–1,500	8
148		*L. platynum*		*Collemataceae*			800–1,500	8
149		*L. trichophorum*		*Collemataceae*			800–1,500	8
150	32	*Stereocaulon coniophyllum*	6	*Stereocaulaceae*	1	12	3,300–3,600	4
151		*S. foliolosum*		*Stereocaulaceae*			2,400–4,000	17
152		*S. foliolosum var. botryophorum*		*Stereocaulaceae*			3,600–3,900	4
153		*S. foliolosum var. strictum*		*Stereocaulaceae*			2,400–3,600	13
154		*S. glareosum*		*Stereocaulaceae*			4,200–4,400	3
155		*S. himalayense*		*Stereocaulaceae*			2,550–5,400	29
156		*S. myriocarpum*		*Stereocaulaceae*			2,700–3,000	4
157		*S. paradoxum*		*Stereocaulaceae*			3,450–3,900	5
158		*S. piluliferum*		*Stereocaulaceae*			1,800–3,900	22
159		*S. pomiferum*		*Stereocaulaceae*			2,500–4,700	23
160		*S. sasaki var. sasaki*		*Stereocaulaceae*			3,600–4,200	7
161		*S. sasakii*		*Stereocaulaceae*			3,600–4,200	7
162	33	*Lobaria isidiosa*	7	*Lobariaceae*	2	9	1,900–3,700	19
163		*L. kurokawae*		*Lobariaceae*			1,800–3,400	17
164		*L. pseudopulmonaria*		*Lobariaceae*			2,550–4,050	16
165		*L. retigera*		*Lobariaceae*			1,600–3,650	22
166	34	*Sticta cyphellulata*		*Lobariaceae*			500–2,100	17
167		*S. limbata*		*Lobariaceae*			1,800–2,420	7

Table 3.1 (continued)

S. No. (Species)	S. No. (Genera)	Terricolous lichen species	S. No. (Family)	Family	No. of genera/family	No. of spp./family	Altitudinal range (m)	Frequency
168		*S. nylanderiana*		*Lobariaceae*			1,800–3,600	19
169		*S. orbicularis*		*Lobariaceae*			1,800–2,400	7
170		*S. weigelii*		*Lobariaceae*			8,00–2,250	16
171	35	*Catapyrenium cinereum*	8	*Verrucariaceae*	1	4	4,300–5,080	9
172		*C. cinereum*		*Verrucariaceae*			4,300–5,080	9
173		*C. daedalium*		*Verrucariaceae*			3,900–6,000	22
174	36	*Placidium squamulosum*		*Verrucariaceae*			4,400–4,500	2
175	37	*Lecanora amorpha*	9	*Lecanoraceae*	1	4	5,000	1
176		*L. chondroderma*		*Lecanoraceae*			3,600–5,600	21
177		*L. himalayae*		*Lecanoraceae*			4,000–5,120	12
178		*L. terestiuscula*		*Lecanoraceae*			4,500–5,200	8
179	38	*Leprocaulon arbuscula*	10	*Imperfect fungi*	1	4	2,700–3,100	5
180	39	*Thamnolia vermicularis*		*Imperfect fungi*			5,455	1
181		*T. vermicularis var. subuliformis*		*Imperfect fungi*			3,600–5,400	19
182		*T. vermicularis var. vermicularis*		*Imperfect fungi*			3,600–5,400	19
183	40	*Nephroma expallidum*	11	*Nephromataceae*	1	4	3,000	1
184		*N. helveticum var. helveticum*		*Nephromataceae*			1,600–3,600	21
185		*N. isidiosum*		*Nephromataceae*			3,300–3,500	3
186		*N. parile*		*Nephromataceae*			2,500–3,600	12
187	41	*Baeomyces pachypus*	12	*Baeomycetaceae*	1	3	1,800–3,600	19
188		*B. roseus*		*Baeomycetaceae*			2000–2500	6
189		*B. soredifer*		*Baeomycetaceae*			2,500	1
190	42	*Coccocarpia erythroxyli*	13	*Coccocarpiaceae*	1	3	2,850	1
191		*C. palmicola*		*Coccocarpiaceae*			800–1,500	8

Table 3.1 (continued)

S. No. (Species)	S. No. (Genera)	Terricolous lichen species	S. No. (Family)	Family	No. of genera/ family	No. of spp./ family	Altitudinal range (m)	Frequency
192		C. pellita		Coccocarpiaceae			800–1,500	8
193	43	Diploschistes muscorum	14	Thelotremataceae	1	3	900–2,160	14
194		D. muscorum subsp. muscorum		Thelotremataceae			2,160	1
195		D. nepalensis		Thelotremataceae			900	1
196	44	Fuscopannaria saltuensis	15	Pannariaceae	2	3	3,100	1
197		F. siamensis		Pannariaceae			2,400–2,600	3
198	45	Gymnoderma coccocarpum		Pannariaceae			3,000–3,300	4
199	46	Ramalina cfr taitensis	16	Ramalinaceae	1	3	1,500–2,500	11
200		R. hossei		Ramalinaceae			1,200–2,500	14
201		R. intermedia		Ramalinaceae			3,800	1
202	47	Dibaeis baeomyces	17	Icmadophilaceae	1	2	1,500–3,600	22
203		D. sorediata		Icmadophilaceae			2,500	1
204	48	Siphula ceratites	18	Siphulaceae	1	2	2,500–3,600	12
205		S. ceratites var. himalayensis		Siphulaceae			4,000	1
206	49	Usnea maculata	19	Usneaceae	1	2	2,000–2,600	7
207		U. subfloridana		Usneaceae			2,200–3,700	16
208	50	Acroscyphus sphaerophoroides	20	Caliciaceae	1	1	4,000–4,400	5
209	51	Alectoria ochroleuca	21	Alectoriaceae	1	1	4,000–5,100	12
210	52	Catolechia wahlenbergii	22	Rhizocarpaceae	1	1	4,500	1
211	53	Mycobilimbia hunana	23	Porpidiaceae	1	1	1,650	1
212	54	Psora himalayana	24	Psoraceae	1	1	1,200	1

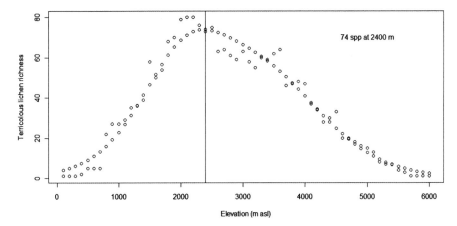

Fig. 3.1 Altitudinal richness pattern of soil lichens. The fitted regression line represents the statistically significant ($P \leq 0.001$) smooth spline (s) after using *GAM* with approximately 4 degrees of freedom. The *vertical line* in the space represents the highest richness

Table 3.2 Regression analysis results modelled after different terricolous lichen species richness as well as their six dominant families as response variables and their elevation as predictor variable. The quasi-Poisson family of error fitted in the GAM model after the cubic smooth spline (s) with approximately 4 degrees of freedom and the total degrees of freedom in this observation are 59

Response variables	Resid. df	Res. dev	D^2	Deviance	F	Pr (>F)
Total soil lichens	55	57	0.966	1622	448	<0.0001
Cladoniaceae	55	38	0.958	856.7	461.87	<0.0001
Parmeliaceae	55	33.59	0.903	311.67	163.54	<0.0001
Peltigeraceae	55	18.33	0.940	285.5	304.37	<0.0001
Physciaceae	55	40.11	0.864	255	103.72	<0.0001
Collemataceae	55	39.86	0.794	153.78	60.68	<0.0001
Stereocaulaceae	55	13	0.936	190.19	247.07	<0.0001

Resid. df Residual degree of freedom, *Res. dev* Residual Deviance, D^2 Regression coefficient of determination, *F* Fisher value of determination, *Pr* the Probability

The elevation for the highest richness of soil lichens represents the temperate forest zone within the Himalayas. Temperate vegetation occurred at 2,400 m altitude dominated by deciduous tree species such as *Schima-Castanopsis*, laurels, oaks (*Quercus* spp.) and evergreen species such as blue pine (*Pinus wallichiana*). The larger vascular species may provide shade and has relatively higher humidity due to their closed canopy. The climate of this zone is characterized by moderate temperature throughout the whole year and relatively higher precipitation and evapotranspiration than elsewhere. The climatic conditions in the Himalayan temperate habitats help to maintain greater heterogeneity among microhabitats and enough

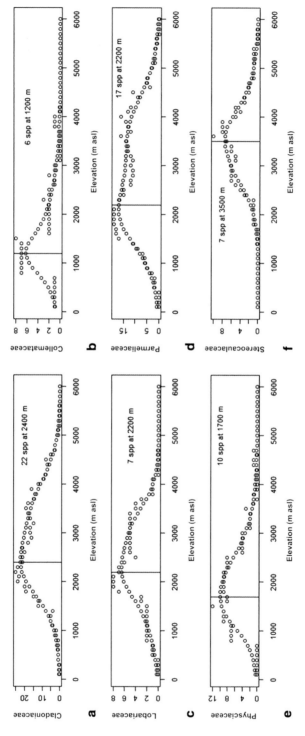

Fig. 3.2 Altitudinal richness pattern showed by dominant soil lichen families. **a** *Cladoniaceae*; **b** *Parmeliaceae*; **c** *Peltigeraceae*; **d** *Physciaceae*; **e** *Collemataceae* and **f** *Stereocaulaceae*. The *fitted regression line* represents the statistically significant ($P \leq 0.001$) smooth spline (s) after using GAM with approximately 4 degrees of freedom. The *vertical line* in each figure represents the highest richness

soil moisture which supports good growth of vascular plants and soil lichens. Thus higher soil lichen distribution is expected in these temperate habitats, nurtured by moderate regimes of temperature, precipitation and solar radiation. This view is well supported by development and distribution of soil lichen communities favoured by occurrence of vascular plants in the Great Basin, western USA (Clair et al. 1993). But it differed from the findings of Baniya et al. (2010) and Rai et al. (2011, 2012), which recorded the highest richness from 3,000–3,400 m. These differences may be the result of higher zooanthropogenic perturbations at lower altitudes (≤2,400 m) and consequent shifting of soil lichens to mid-elevations (i.e., 3,000–3,400 m).

Diversity of microhabitats in the form of "doons" and valleys and alpine pastures—*Bugyals* along with a thinning treeline between the altitudinal ranges of 2,000–3,000 m provide ideal conditions for soil lichen communities to develop and flourish. Maximum Bugyals in the Himalayas are situated at average altitude of 2,900–3,400 m, where due to thinning out of treeline, there is least competition with angiospermic vegetation and the harsh climatic regimes (low pressure, high wind velocity, subzero atmospheric temperature, and soil acidity) further increase the probability of soil lichens to colonise, as Himalayan soil lichen species are tolerant to climate extremes (Scheidegger and Clerc 2002). Beside these, majority of Himalayan habitats are now a part of protected area networks, which along with other plant diversity also help soil lichens to colonise and expand.

The number of soil lichens (212) is much higher than earlier studies such as 48 species found in arid and semiarid Australia (Eldridge 1996), 28 species in the northern Namib Desert (Lalley and Viles 2005), 31 species from the southern Namib Desert (Jürgens and Niebel-Lohmann, 1995), 34 species from intermountains of western United States (Clair et al. 1993), 45 species in the Nepal Himalayas (Baniya et al. 2010), and 20 species in Chopta-Tungnath tract in the Garhwal Himalaya (Rai et al. 2011, 2012). The higher species count of soil lichens in Himalayan habitats is due to low competition from other ground vegetation above treeline (≥3000 m), tolerance of specific fruticose/dimorphic terricolous lichens (e.g., *Stereocaulon, Cladonia*) to grazing induced trampling and harsh climatic conditions, sufficient hydration of soils throughout the year, tolerance to acidic pH of soil and efficient utilisation of soil macrohabitats (Grabherr 1982; Baniya et al. 2010; Rai et al. 2012; Rai 2012). Change in landuse patterns, anthropogenic pressures, livestock grazing and ground soil cover along with altitudinal gradients are also major factors influencing the terricolous lichen diversity in Himalayas (Rai et al. 2011, 2012).

Anthropogenic pressures, after the increase in human population, is the main controlling driver of nature and its biodiversity on Earth (Ellis and Ramankutty 2008; Ellis 2011). Major concentration of human population can generally be expected at planes or flat areas between the undulating topography of the Himalayas. Dense human population is mostly found in and around the tropical and subtropical range of the Himalayas (i.e., foothills) and less at or above 3,000 m. The anthropogenic pressure arising by commercial aspects of human society has been a major contributor of overall depletion of lichen diversity in Himalayas (Upreti et al. 2005). Many terricolous lichen species (i.e., *Peltigera, Thamnolia, Cladonia, Staereocaulon*) have been exploited as supplementary foods and medicine by different ethnic groups as well as other animals (Saklani and Upreti 1992; Upreti and Negi 1996; Negi 2003; Upreti et al. 2005; Wang et al. 2001).

8 Conclusion

Elevation is the main driver of biological diversity which acts as a composite factor to the anthropogenic pressures, historical location, and higher plant biodiversity. The terricolous lichen diversity in the Himalayas is modulated by natural factors, e.g., climate, topography, soil chemistry, comparative soil cover along altitude, suitable micro habitats, decrease competition; other ground vegetation are also highly guided by zooanthropogenic pressure, e.g., livestock grazing, commercial/ethnic exploitation, tourism etc. Mid-elevation range of 1,200–3,600 m in temperate habitats of the Himalayas are best suitable for soil lichen to make sustainable communities and can be the most suitable thrust area for research in the future.

References

Awasthi DD (2007) A compendium of the macrolichens from India, Nepal and Sri Lanka. Bishen Singh andMahendra Pal Singh, Dehra Dun

Baniya CB, Solhøy T, Gauslaa Y, Palmer MW (2010) The elevation gradient of lichen species richness in Nepal. Lichenologist 42:83–96

Belnap J, Eldridge DJ (2001) Disturbance and recovery of biological soil crusts. In: Belnap J, Lange OL (eds) Biological soil crusts: structure, function, and management [Ecological studies, Vol. 150]. Springer, Berlin, pp 363–383

Belnap J, Gillette DA (1998) Vulnerability of desert biological soil crusts to wind erosion: the influence of crust development, soil texture and disturbance. J Arid Environ 39:133–142

Belnap J, Büdel B, Lange OL (2001a) Biological soil crusts: characteristics and distribution. In: Brlnap J, Lange OL (eds) Biological soil crusts: structure, function, and management [Ecological studies, Vol. 150]. Springer, Berlin, pp 3–30

Belnap J, Eldridge DJ, Kaltenecker JH, Rosentreter R, Williams J, Leonard S (2001b) Biological soil crusts: ecology and management. U.S. Department of the Interior, Bureau of Land Management, U.S. Geological Survey, Technical Reference. 1730–2:1–110

Beymer RJ, Klopatek, JM (1992) The effects of grazing on cryptogamic crusts in Piyon- Juniper woodlands in Grand Canyon National Park. Am Midl Nat 127:139–148

Clair LL, Johansen JR, Clair SB, Knight KB (2007) The Influence of Grazing and Other Environmental Factors on Lichen Community Structure along an Alpine Tundra Ridge in the Uinta Mountains, Utah, U.S.A. Arct Antarct Alp Res 39:603–613

Clair LS, Johansen JR, Rushforth SR (1993).Lichens of soil crust communities in the intermountain area of the western United States. Gt Basin Nat 3:5–12

Crawley MJ (2006). Statistics: an introduction using R. London: Wiley

Eldridge DJ (1996) Distribution and floristics of lichens in soil crusts in arid and semiarid New South Wales, Australia. Aust J Bot 44:581–599

Eldridge DJ (1998) Trampling of microphytic crusts on calcareous soils and its impact on erosion under rain- impacted flow. Catena 33:221–239

Eldridge DJ, Leys JF (2003) Exploring some relationships between biological soil crusts, soil aggregation and wind erosion. J Arid Environ 53:457–466

Eldridge DJ, Tozer ME (1997) Environmental factors relating to the distribution of terricolous bryophytes and lichens in semi-arid eastern Australia. Bryologist 100:28–39

Ellis EC (2011) Anthopogenic transformation of the terrestrial biosphere. Proc Royal Soc A: Math Physi Eng Sci 369:1010–1035

Ellis EC, Ramankutty N (2008) Putting people in the map: anthropogenic biomes of the world. Front Ecol Environ 6:439–447

Escurado A, Martínez I, Cruz de la A, Otálora MAG, Maestre FT (2007) Soil lichens have species-specific effects on the seedling emergence of three gypsophile plant species. J Arid Environ 70:18–28

Evans RD, Belnap J (1999) Long-term consequences of disturbance on nitrogen dynamics in an arid ecosystem. Ecology 80:150–160

Favero-Longo SE, Piervittori R (2010) Lichen-plant interactions. J Plant Interact 5:163–177

Gardner CR, Mueller DMJ (1981) Factors affecting the toxicity of several lichen: effect of pH and lichen acid concentration. Am J Bot 68:87–95

Grabherr G (1982) The impact of trampling by tourists on a high altitudinal grassland in the Tyro-lean Alps, Austria. Vegetatio 48:209–217

Harper KT, Belnap J (2001) The influence of biological soil crusts on mineral uptake by associated vascular plants. J Arid Environ 47(3):347–357

Hastie TJ, Tibshirani RJ (1990) Generalised additive models. London: Chapman and Hall.

Heegaard E (2004) Trends in aquatic macrophyte species turnover in Northern Ireland – which factors determine the spatial distribution of local species turnover? Glob Ecol Biogeogr 13:397–408

Huang MR (2010) Altitudinal patterns of *Stereocaulon* (Lichenized Ascomycota) in China. Acta Oecologica 36:173–178

Jürgens N, Niebel-Lohmann A (1995) Geobotanical observations on lichen fields of the southern Namib Desert. Mitteilungen aus dem Institut für Allgemeine Botanik in Hamburg 25:135–156

Lalley JS, Viles HA, Copeman, N, Cowley C (2006a) The influence of multi-scale variables on the distribution of terricolous lichens in a fog desert. J Veg Sci 17:831–838

Lalley JS, Viles HA (2005) Terricolous lichens in the northern Namib Desert of Namibia: distribu-tion and community composition. Lichenologist 37:77–91

Lalley JS, Viles HA, Henschel JR, Lalley V (2006b) Lichen-dominated soil crusts as arthropod habitat in warm deserts. J Arid Environ 67:579–593

Lawrey JD (1977) Adaptive significance of O-methylated lichen depsides and depsidones. Liche-nologist 9:137–142

Lawrey JD (2009) Diversity of defensive mutualisms. In: Chemical defense in lichen symbiosis. Taylor and Francis Group. London, pp 167–181

Motiejûnaitë J, Faùtynowicz W (2005) Effect of land-use on lichen diversity in the transboundary region of Lithuania and northeastern Poland. Ekologija 3:34–43

Negi HR (2003) Lichens: a valuable bioresource for environmental monitoring and sustainable development. Resonance 8:51–58

Negi HR, Gadgil M (1996) Patterns of distribution of Macrolichens in western parts of Nanda Devi Biosphere Reserve. Curr Sci 71:568–575

Pellant M, Shaver P, Pyke DA, Herrick JE (2001) Interpreting indicators of rangeland health, TR-1734-5, US Dept. of the interior, Denver, Colorado

Pinokiyo A, Singh KP, Singh JS (2008) Diversity and distribution of lichens in relation to altitude within a protected biodiversity hot spot, north-east India. Lichenologist 40:47–62

R Development Core Team (2011) R: A language and environment for statistical computing ver-sion R 2.13.1

Rai H (2012) Studies on diversity of terricolous lichens of Garhwal Himalaya with special refe-rence to their role in soil stability. PhD Thesis. H.N.B Garhwal University. Srinagar (Garhwal), Uttarakhand, India

Rai H, Nag P, Upreti DK, Gupta RK (2010) Climate Warming Studies in Alpine Habitats of Indian Himalaya, using Lichen based Passive Temperature-enhancing System. Nat Sci 8:104–106

Rai H, Khare R, Gupta RK, Upreti DK (2011) Terricolous lichens as indicator of anthropogenic disturbances in a high altitude grassland in Garhwal (Western Himalaya), India. Botanica Orientalis: J Plant Sci 8:16–23

Rai H, Upreti DK, Gupta RK (2012) Diversity and distribution of terricolous lichens as indicator of habitat heterogeneity and grazing induced trampling in a temperate-alpine shrub and meadow. Biodivers Conserv 21:97–113

Saklani A, Upreti DK (1992) Folk uses of some lichens in Sikkim. J Ethnopharmacol 37:229–233

Scheidegger C, Clerc P (2002) Rote Liste der gefährdeten Arten der Schweiz: Baum- und erdbewohnende Flechten. Hrsg. Bundesamt für Umwelt, Wald und Landschaft BUWAL, Bern, und Eidgenössische Forschungsanstalt WSL, Birmensdorf, und Conservatoire et Jardin botaniques de la Ville de Genève CJBG. BUWAL-Reihe Vollzug Umwelt. pp 124

Scutari NC, Bertiller MB, Carrera AL (2004) Soil-associated lichens in rangelands of north-eastern Patagonia. Lichen groups and species with potential as bioindicators of grazing disturbance. Lichenologist 36:405–412

Sedia EG, Ehrenfeld JG (2005) Differential effects of lichens, mosses and grasses on respiration and nitrogen mineralization in soils of the New Jersey Pinelands. Oecologia 144:137–147

Sharma LR (1995) Enumeration of lichens of Nepal. Biodiversity Profiles Project Euroconsult (Publication No. 3)

Singh JS, Singh SP (1987) Forest vegetation of the Himalaya. Bot Rev 53:80–192

Upreti DK (1998) Diversity of lichens in India. In: Agarwal SK, Kaushik JP, Kaul KK, Jain AK (eds) Perspectives in Environment. APH Publishing Corporation, New Delhi, pp 71–79

Upreti DK, Negi HR (1996) Folk use of *Thamnolia vermicularis* (Swartz.) Ach. in Schaerer in Lata vellage of Nanda Devi Biosphere Reserve. Ethnobotany 8:83–86

Upreti DK, Divakar PK, Nayaka S (2005) Commercial and ethnic use of lichens in India. Econ Bot 59:269–273

Wang LS, Narui T, Harada H, Culberson CF, Culberson WL (2001) Ethnic uses of lichens in Yunnan, China. Bryologist 104:345–349

Whittaker RH (1972) Evolution and measurement of species diversity. Taxon 21:213–251

Will-Wolf S, Esseen PA, Neitlich P (2002) Methods for monitoring biodiversity and ecosystem function. In: Nimis PL, Scheidegger C, Wolseley PA (eds) Monitoring with lichens-monitoring lichens. [NATO Science Series IV: earth and environmental science vol. 7]. Kluwer Academic Publishers, Dordrecht, p 147–162

Zaady E, Bouskila A (2002) Lizard burrows association with successional stages of biological soil crusts in an arid sandy region. J Arid Environ 50:235–246

Chapter 4
Photobiont Diversity in Indian *Cladonia* Lichens, with Special Emphasis on the Geographical Patterns

Tereza Řídká, Ondřej Peksa, Himanshu Rai, Dalip Kumar Upreti and Pavel Škaloud

1 Introduction

Lichens show distinctive patterns of distribution at both micro and macro levels (Galloway 2008). Sixteen major biogeographical patterns have been distinguished in lichens, including cosmopolitan taxa, bipolar taxa, taxa specific for particular continents or areas, and endemic taxa (Galloway 2008). However, these patterns are applicable to lichen-forming fungi only. Till date, we have almost no idea about the biogeography of lichenized algae and cyanobacteria—the photobionts.

During the last 20 years, molecular phylogenetic studies dramatically changed our views regarding coevolution of lichen partners. Supposed cospeciation and parallel cladogenesis of mycobionts and photobionts has been generally rejected (Kroken and Taylor 2000; Piercey-Normore and DePriest 2001), and replaced with the domestication model, in which the fungal partner could select the best available photobiont (DePriest 2004). In general, the mycobionts are able to cooperate with several algal species and to switch them (Muggia et al. 2008; Nelsen and Gargas 2009; Nyati 2007; Piercey-Normore 2006; Wornik and Grube 2010), simultaneously, several mycobionts can share single algal partner (Beck 1999; Doering and Piercey-Normore 2009; Hauck et al. 2007; Piercey-Normore 2009; Rikkinen et al. 2002). Moreover, lichen algae and cyanobacteria could exhibit their own environmental requirements, which seem to be independent of particular mycobionts to a large extent (Cordeiro et al. 2005; Fernandez-Mendoza et al. 2011;

P. Škaloud (✉) · T. Řídká
Department of Botany, Faculty of Science, Charles University,
Benátská 2, 12801, Prague, Czech Republic
e-mail: skaloud@natur.cuni.cz

O. Peksa
The West Bohemian Museum in Pilsen, Plzeň, Czech Republic

H. Rai · D. K. Upreti
Lichenology laboratory, Plant Diversity, Systematics and Herbarium Division
CSIR-National Botanical Research Institute,
Rana Pratap Marg, Lucknow
Uttar Pradesh-226001, INDIA

H. Rai, D. K. Upreti (eds.), *Terricolous Lichens in India,*
DOI 10.1007/978-1-4614-8736-4_4, © Springer Science+Business Media New York 2014

Helms 2003; Muggia et al. 2008; Peksa and Škaloud 2011). Naturally, since the environmental preferences of an organism could be narrowly linked to its distribution, the geographical pattern of photobionts could be markedly different from that of their fungal partners.

Both cyanobacteria and algae are microscopic organisms. Moreover, the lichen vegetative propagules containing both symbionts (soredia, isidia, etc.) are mostly not much larger than particular algal cells (20–50 µm, Büdel and Scheidegger 2008) and they are capable of being dispersed over large distances as well (Bailey 1976). The well-known theory of ubiquitous dispersal of microbial species (Finlay and Clarke 1999) presumed that most organisms smaller then ca 1 mm should occur worldwide (in a niche-based context only).

Indeed, some photobiont lineages are apparently widely distributed. For example, *Asterochloris* clade A7 (sensu Peksa and Škaloud 2011) has been found in lichens collected from Europe, USA, and China, indicating its ubiquitous dispersal. On the other hand, many photobiont lineages have been reported only from specific continents or regions. However, because of very uneven distribution of lichen collections, it is premature to classify them as species with narrow distribution patterns. For the most studied photobiont genera (*Asterochloris*, *Trebouxia*, *Nostoc*), majority of reports have been published from Europe and North America (e.g., Bačkor et al. 2010; Blaha et al. 2006; Guzow-Krzemińska 2006; Nelsen and Gargas 2008; O'Brien et al. 2005; Paulsrud et al. 2000; Peksa and Škaloud 2011; Piercey-Normore 2004, 2006, 2009; Yahr et al. 2004), slightly less from Central and South America (Cordeiro et al. 2005; Helms 2003; Reis 2005) and Antarctica (Aoki et al. 1998; Engelen et al. 2010; Nyati 2007; Otálora et al. 2010; Romeike et al. 2002; Wirtz et al. 2003). However, only few or no data have been reported from Africa, Asia, Australia, and close islands (Helms 2003; Nelsen and Gargas 2008, 2009; Nyati 2007; Piercey-Normore and DePriest 2001). Therefore, further exploration of photobionts in these areas is necessary to obtain relevant information about biogeographical patterns in lichenized algae and cyanobacteria.

2   Objectives

In this study, we investigated *Asterochloris* photobionts from terricolous lichens (*Cladonia* spp.) collected in India and Nepal using DNA sequencing. Traditionally, *Asterochloris* (incl. former *Trebouxia*) species have been determined according to the morphological features such as cell shape, chloroplast structure, and pyrenoid ultrastructure. However, a large cryptic variability recently discovered within the genus (Piercey-Normore and DePriest 2001; Yahr et al. 2004; Škaloud and Peksa 2010) clearly points out the deficiency of morphological features to delimit real species entities within *Asterochloris*. Therefore, we sequenced the internal transcribed spacer (ITS) ribosomal DNA (rDNA) marker to genetically investigate the diversity of photobionts in *Cladonia* lichens. The newly obtained ITS rDNA sequences were added to the dataset of all sequences deposited in GenBank database to analyze the phylogenetic position of Indian photobionts and biogeographic patterns of selected *Asterochloris* lineages.

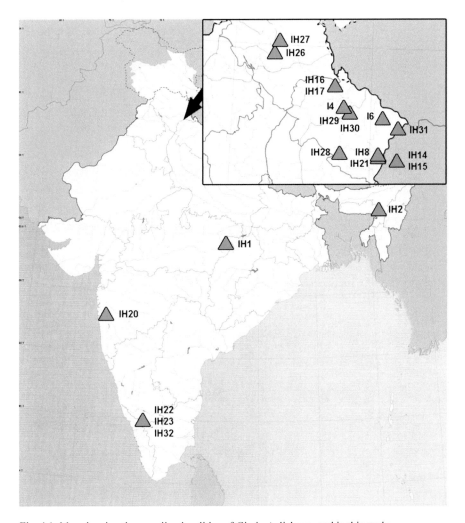

Fig. 4.1 Map showing the sampling localities of *Cladonia* lichens used in this study

3   Materials and Methods

3.1   Sample Collection

Lichen samples were collected in five different areas in India and Nepal (Fig. 4.1). Single lichen sample was collected in Maharashtra, Madhya Pradesh, and Assam states, located in west, central, and north-eastern India, respectively. Three lichen samples were obtained from collections made in Tamil Nadu state, located in South India. The majority of lichen thalli were collected in the Himalayas, Uttarakhand, and Himachal Pradesh states. Finally, two lichen thalli were collected in eastern Nepal. The collections have been made at different times during the years 2007 and 2010 (Table 4.1).

Table 4.1 List of *Cladonia* samples used for sequencing of algal internal transcribed spacer (ITS) ribosomal DNA (rDNA)

Sample No.	Taxon	Date of Collection	Altitude (m)	Country	State	District	Site	Substratum	Lattitude	Longitude
I4	*Cladonia furcata* (Huds.) Schrad.	23.8.2007	3,250	India	Uttarakhand	Rudra-prayag	Tungnath	Soil	30°29'18.9"N	79° 12' 54.4"E
I6	*Cladonia rangiferina* (L.) Weber ex Wigg.	18.10.2007	2,553	India	Uttarakhand	Pithoragarh	Between Bog-diyar and Naher Devi	Soil	30°13'32.9"N	80°13'11.5"E
IH1	*Cladonia praetermissa* A.W. Archer	10.5.2010	1,665	India	Madhya Pradesh	Anuppur	Shambhudhara, Amarkantak Protected area	Red hard soil	22°42'48.9"N	81°44'46.3"E
IH2	*Cladonia scabriuscula* (Delise) Nyl.	18.11.2008	1,014	India	Assam	North Cachar	Haflong	Soil	25°08'01.3"N	93°00'35.5"E
IH8	*Cladonia verticillata* (Hoffm.) Schaer.	29.10.2009	1,890	India	Uttarakhand	Champawat	Lohaghat to Mayawati	Soil	29°23'54.5"N	80°05'07.1"E
IH14	*Cladonia coniocraea* (Flörke) Spreng.	18.5.2010	1,800	Nepal	Mahakali zone	Dadeldhura	Dadeldhura community forest	Soil in biological soil crust	29°18'08.5"N	80°35'32.7"E

Table 4.1 (continued)

Sample No.	Taxon	Date of Collection	Altitude (m)	Country	State	District	Site	Substratum	Lattitude	Longitude
IH15	*Cladonia coniocraea* (Flörke) Spreng.	17.5.2010	1,716	Nepal	Mahakali zone	Dadeldhura	Dadeldhura community forest	Bark of Rhodo- dendron arboreum	29°18'07.1"N	80°35'31.4"E
IH16	*Cladonia pyxidata* (L.) Hoffm.	29.10.2010	3,118	India	Uttarakhand	Uttarkashi	Gangotri	Soil	30°59'34.3"N	78°56'21.2"E
IH17	*Cladonia fruiticulosa* Kremp.	29.10.2010	3,100	India	Uttarakhand	Uttarkashi	Gangotri	Soil	30°59'35.3"N	78°56'20.5"E
IH20	*Cladonia scabriuscula* (Delise) Nyl.	26.3.2010	1,410	India	Maharashtra	Satara	Mahabalesh- war, Wilson Point	Rocks with mosses	17°55'17.6"N	73°40'23.4"E
IH21	*Cladonia dela- vayi* Abbayes	27.11.2010	1,618	India	Uttarakhand	Champawat	Marodkhan on way to Ghat	Rock	29°19'57.7"N	80°05'27.8"E
IH22	*Cladonia fruiticulosa* Kremp.	12.1.2008	2,607	India	Tamil Nadu	Nilgiri	Dodabetta, trails from Samer to Tiger Hills	Soil in coni- ferous forest	11°23'45.4"N	76°43'36.6"E
IH23	*Cladonia fur- cata* (Huds.) Schrad.	12.1.2008	2,607	India	Tamil Nadu	Nilgiri	Dodabetta, trails from Samer to Tiger Hills	Soil in coni- ferous forest	11°23'45.4"N	76°43'36.6"E
IH26	*Cladonia fur- cata* (Huds.) Schrad.	5.6.2008	3,078	India	Himachal Pradesh	Kullu	On route to Dhela	Soil	31°42'17.2"N	77°16'14.6"E

Table 4.1 (continued)

Sample No.	Taxon	Date of Collection	Altitude (m)	Country	State	District	Site	Substratum	Lattitude	Longitude
IH27	Cladonia furcata (Huds.) Schrad.	4.5.2008	2,300	India	Himachal Pradesh	Kullu	7 km before Pulga	Soil among mosses	31°59'53.3"N	77°24'47.4"E
IH28	Cladonia cariosa (Ach.) Spreng.	29.10.2009	1,745	India	Uttarakhand	Champawat	Mayawati to Lohaghat	Rock	29°25'22.6"N	79°04'29.7"E
IH29	Cladonia pyxidata (L.) Hoffm.	8.6.2008	3,550	India	Uttarakhand	Chamoli	Kothidhar	Soil	30°23'34.9"N	79°19'08.7"E
IH30	Cladonia furcata (Huds.) Schrad.	8.6.2008	3,700	India	Uttarakhand	Chamoli	Srenikhal	Soil	30°22'27.5"N	79°19'10.3"E
IH31	Cladonia corymbescens Nyl. ex Leight	2.11.2009	2,743	India	Uttarakhand	Pithoragarh	Narain Swami Ashram	Soil	29°58'15.3"N	80°39'19.7"E
IH32	Cladonia fruticulosa Kremp.	12.1.2008	2,607	India	Tamil Nadu	Nilgiri	Dodabetta, trails from Samer to Tiger Hills	Soil in coniferous forest	11°24'00.9"N	76°44'06.2"E

3.2 DNA Isolation, Polymerase Chain Reaction (PCR), and Sequencing

Total genomic DNA was extracted following the standard CTAB protocol (Doyle and Doyle 1987), with minor modifications. The total genomic DNA was dissolved in sterile dH_2O and amplified by polymerase chain reaction (PCR). The ITS1-5.8S-ITS2 rDNA region was amplified using universal primer ITS4-3' (5'-TCCTCCGCT-TATTGATATGC-3'; White et al. 1990) and the algal-specific primer nr-SSU-1780-5' (5'-CTGCGGAAGGATCATTGATTC-3'; Piercey-Normore and DePriest 2001). All PCR reactions were performed in total volume of 20 µl contained 12.4 µl of sterile Mili-Q water, 2 µl of AmpliTaq Gold® 360 Buffer 10× (Applied Biosystems, Life technologies, Carlsbad, CA, USA), 1.5 µl of $MgCl_2$ (25 mM), 0.4 µl of dNTP mix (10 mM), 0.25 µl of each primer (25 nM), 2 µl of 360 GC Enhancer, 0.2 µl of AmpliTaq Gold® 360 DNA Polymerase and 1 µl of DNA (10 ng·l^{-1}). PCR and cycle-sequencing reactions were performed in a Touchgene Gradient cycler (Krackeler Scientific, Albany, NY, USA). PCR amplification of the algal ITS rDNA began with an initial denaturation at 95 °C for 10 min, followed by 35 cycles of denaturing at 95 °C for 1 min, annealing at 50 °C for 1 min and elongation at 72 °C for 1 min, with a final extension at 72 °C for 10 min. The PCR products were quantified on a 1 % agarose gel stained with ethidium bromide and purified using the JetQuick PCR Purification kit (Genomed, Löhne, Germany), according to the manufacturer's protocols. The purified amplification products were sequenced with PCR primers using an Applied Biosystems (Seoul, Korea) automated sequencer (ABI 3730xl) at Macrogen Corp. in Seoul, Korea. Sequencing reads were assembled and edited using the SeqAssem programme (Hepperle 2004).

4 Phylogenetic Analyses

The newly obtained ITS rDNA sequences were added to the concatenated (ITS rDNA, actin I locus) alignment analyzed in Škaloud and Peksa (2010). Then, we added several additional ITS rDNA sequences obtained from GenBank to cover all *Asterochloris* diversity. The final concatenated matrix containing 69 taxa, was 1137 bp long, and was 100 % filled for the ITS data and 67 % filled for the actin data (Table 4.2). The matrix is available from Pavel Škaloud The phylogenetic tree was inferred with Bayesian inference (BI) using MrBayes version 3.1 (Ronquist and Huelsenbeck 2003). The analysis was carried out on the partitioned dataset using the strategy described in Peksa and Škaloud (2011). Bootstrap analyses were performed by maximum likelihood (ML) and weighted parsimony (wMP) criteria using GARLI, version 0.951 (Zwickl, 2006) and PAUP version 4.0b10 (Swofford 2002), respectively. ML analysis consisted of rapid heuristic searches (100 pseudo-replicates) using automatic termination (genthreshforto-poterm command set to 100,000). The wMP bootstrapping (1,000 replications)

Table 4.2 List of all samples used in the study, including GenBank accession numbers for photobiont sequences. The samples are ordered with respect to their position in the Bayesian phylogenetic tree (Fig. 4.2)

Clade No.	Fungal taxon	Origin	Collection number	GenBank accession	
				ITS	Actin
1	*Cladonia squamosa* (Scop.) Hoffm.	USA, MA	CAUP H1006	AF345406	AM906025
1	*Stereocaulon pileatum* Ach.	USA, MA	UTEX 896	AF345404	–a
1	*Stereocaulon pileatum* Ach.	USA, MA	UTEX 897	AF345405	–a
1	*Stereocaulon pileatum* Ach.	USA, MA	UTEX 1713	AF345407	–a
1	*Stereocaulon* sp.	Slovakia	Peksa 801	FM945392	–a
1	*Stereocaulon evolutoides* (H. Magn.) Frey	USA, MA	UTEX 895	AF345382	AM906024
1	*Cladonia coniocraea* (Flörke) Spreng.	Nepal	IH15	HE803028b	–
1	*Stereocaulon vesuvianum* Pers.	USA, Alaska	Talbot 281	DQ229885	DQ229888
2	*Stereocaulon botryosum* Ach.	USA, Alaska	Talbot 153	DQ229880	DQ229889
2	*Stereocaulon pileatum* Ach.	Czech Republic	Peksa 999	AM905999	AM906028
2	*Stereocaulon subcoralloides* (Nyl.) Nyl.	USA, Alaska	Talbot 167	DQ229881	DQ229890a
2	*Stereocaulon* sp.	Iceland	UTEX 2236	AF345411	AM906027a
3	*Pilophorus aciculare* (Ach.) Th. Fr.	USA, WA	CAUP H1004	AM906012	AM906041
4	*Cladonia cristatella* Tuck.	USA, MA	CAUP H1005	AF345440	AM906018
11	*Cladonia rangiferina* (L.) Weber ex F.H. Wigg.	India	I6	HE803029b	–
11	*Cladonia furcata* (Huds.) Schrad.	India	IH27	HE803030b	–a
11	*Cladonia pyxidata* (L.) Hoffm.	India	IH29	HE803031b	–
11	*Cladonia furcata* (Huds.) Schrad.	India	IH30	HE803032b	–a
11	*Cladonia corymbescens* Nyl. ex Leight	India	IH31	HE803033b	–a
11	*Cladonia corymbescens* Nyl. ex Leight	India	IH31a	HE803034b	–a

Table 4.2 (continued)

Clade No.	Fungal taxon	Origin	Collection number	GenBank accession ITS	Actin
12	*Cladonia furcata* (Huds.) Schrad.	India	14	HE803035b	–
12	*Cladonia furcata* (Huds.) Schrad.	India	IH23	HE803036b	–a
12	*Cladonia furcata* (Huds.) Schrad.	India	IH26	HE803037b	–
5	*Stereocaulon dactylophyllum* Flörke	USA, VT	UTEX 1714	AM905993	AM906019
6	*Lepraria caesioalba* (de Lesd.) J.R.Laundon	Czech Republic	Peksa 173	AM906003	AM906032
6	*Lepraria neglecta* (Nyl.) Lettau	Czech Republic	Peksa 183	AM906002	AM906031a
6	*Lepraria neglecta* (Nyl.) Lettau	Czech Republic	Peksa 207	AM906005	AM906034a
A9	*Lepraria alpina* (de Lesd.) Tretiach & Baruffo	Spain	Peksa 860	FN556035	FN556048
7	*Lepraria rigidula* (de Lesd.) Tønsberg	Czech Republic	Peksa 236	AM905997	AM906023
7	*Lepraria rigidula* (de Lesd.) Tønsberg	Czech Republic	Peksa 900	FM955669	FM955673
7	*Lepraria borealis* Lohtander & Tønsberg	USA, CA	Peksa 869	FN556039	FN556049
7	*Lepraria caesioalba* (de Lesd.) J.R.Laundon	USA, CA	Peksa 873	FN556042	FN556051
7	*Lepraria* sp.	USA, CA	Peksa 870	FN556043	FN556052
7	*Lepraria caesioalba* (de Lesd.) J.R.Laundon	USA, CA	Peksa 872	FN556041	FN556050a
9	*Cladonia evansii* Abbayes	USA, FL	RY798	AY712691	–
9	*Cladonia subtenuis* (Abbayes) Mattick	USA, FL	RY999	AY712698	–
9	*Cladonia subtenuis* (Abbayes) Mattick	USA, FL	RY941	AY712690	–
9	*Cladonia scabriuscula* (Delise) Nyl.	India	IH20	HE803038b	–
9	*Cladonia coniocraea* (Flörke) Spreng.	Nepal	IH14	HE803039b	–

Table 4.2 (continued)

Clade No.	Fungal taxon	Origin	Collection number	GenBank accession	
				ITS	Actin
9	*Cladonia delavayi* Abbayes	India	IH21A	HE803040b	–a
9	*Cladia aggregata* (Sw.) Nyl.	Costa Rica	Nelsen 2138	EU008658	–
9	*Cladonia verticillata* (Hoffm.) Schaer.	India	IH8B	HE803041b	–
9	*Cladonia delavayi* Abbayes	India	IH21B	HE803042b	–
9	*Cladonia scabriuscula* (Delise) Nyl.	India	IH2	HE803043b	–
9	*Cladonia spinea* Ahti	Guyana	MN-069	AF345418	–a
9	*Cladonia variegata* Ahti	Guyana	MN-075	AF345419	–a
9	*Cladonia subtenuis* (Abbayes) Mattick	USA, PA	RY1225	DQ482676	–
9	*Cladonia verticillata* (Hoffm.) Schaer.	India	IH8A	HE803044b	–
9	*Cladonia crinita*	Brazil	-	AY842277	–
9	*Cladonia fruticulosa* Kremp.	India	IH22	HE803045b	–
9	*Cladonia fruticulosa* Kremp.	India	IH32	HE803046b	–a
9	*Cladonia fruticulosa* Kremp.	India	IH32b	HE803047b	–a
9	*Cladonia praetermissa* A.W. Archer	India	IH1	HE803048b	–
9	*Cladonia cariosa* (Ach.) Spreng.	India	IH28	HE803049b	–
9	*Cladonia peltastica* (Nyl.) Muell. Arg.	Guyana	MN-070	AF345416	–
9	*Stereocaulon* sp.	Costa Rica	Nelsen 2181b	DQ229884	DQ229896
9	*Pilophorus* cf. *cereolus* (Ach.) Th. Fr.	Costa Rica	Nelsen 2233f	DQ229883	DQ229895
9	*Lepraria* sp.	Costa Rica	Nelsen L54	EU008684	EU008711

Table 4.2 (continued)

Clade No.	Fungal taxon	Origin	Collection number	GenBank accession	
				ITS	Actin
8	*Cladonia rei* Schaer.	Slovakia	Peksa 787	FM945380	FM955675
8	*Cladonia fimbriata* (L.) Fr	Slovakia	Peksa 796	FM945358	FM955674
10	*Stereocaulon saxatile* H.Magn.	USA, Alaska	Talbot KIS 187	DQ229886	DQ229897
10	*Cladonia foliacea* (Huds.) Willd.	Czech Republic	Peksa 1008	AM906016	AM906049
10	*Lepraria borealis* Lohtander & Tonsberg	Bulgaria	Bayerová 3402	AM906015	AM906048
10	*Lepraria borealis* Lohtander & Tonsberg	Bulgaria	Bayerová 3401	AM900492	AM906045
10	*Lepraria crassissima* (Hue) Lettau	Czech Republic	Peksa 888	FN556033	–
10	*Lepraria yunnaniana* Diederich, Sérus. & Aptroot	Costa Rica	Nelsen 3637b	EU008681	EU008710
11	*Lepraria caesioalba* (de Lesd.) J.R.Laundon	Romania	Peksa 225	AM905996	AM906022
11	*Lepraria caesioalba* (de Lesd.) J.R.Laundon	Slovakia	Peksa 234	AM905994	AM906020
11	*Lepraria lobificans* Nyl.	USA, WI	Nelsen L12	EU008675	EU008704
11	*Lepraria caesioalba* (de Lesd.) J.R.Laundon	Slovakia	Peksa 235	AM905995	AM906021
11	*Lepraria lobificans* Nyl.	USA, WI	Nelsen 154	DQ229877	DQ229898
11	*Lepraria lobificans* Nyl.	USA, WI	Nelsen 153	EU008678	EU008707
11	*Lepraria caesioalba* (de Lesd.) J.R.Laundon	USA, PA	Nelsen L36	EU008664	EU008697
-	*Lepraria* sp.	China	Nelsen L60	EU008690	EU008715
12	*Cladonia rei* Schaer.	Czech Republic	Peksa 921	FM945378	FM955677
12	*Cladonia pocillum* (Ach.) Grognot	Canada	Normore4719	DQ530209	DQ530190a
12	*Cladonia fimbriata* (L.) Fr.	Slovakia	Peksa 815	FM945359	FM955676
12	*Cladonia pyxidata* (L.) Hoffm.	India	IH16	HE803050b	–
12	*Stereocaulon paschale* (L.) Hoffm.	USA, AK	Talbot 101	DQ229887	DQ229891

Table 4.2 (continued)

Clade No.	Fungal taxon	Origin	Collection number	GenBank accession	
				ITS	Actin
13	*Xanthoria parietina* (L.) Th. Fr.	Italy	CCAP 519/5	AM906001	AM906030
13	*Cladonia* sp.	The Netherlands	CAUP H1003	AF345423	DQ229894
A4	*Lepraria caesioalba* (de Lesd.) J.R.Laundon	Czech Republic	Peksa 526	FN556030	–
A4	*Lepraria rigidula* (de Lesd.) Tonsberg	Czech Republic	Peksa 955	FN556032	–a
A4	*Lepraria rigidula* (de Lesd.) Tonsberg	Czech Republic	Peksa 855	FN556031	FN556047
14	*Lepraria caesioalba* (de Lesd.) J.R.Laundon	Czech Republic	Peksa 551	FM955667	FM955671
14	*Lepraria caesioalba* (de Lesd.) J.R.Laundon	Czech Republic	Peksa 185	FM955666	FM955670
14	*Lepraria rigidula* (de Lesd.) Tonsberg	Czech Republic	Peksa 186	AM905992	AM906017
–	*Stereocaulon tomentosum* Fr.	USA, AK	Talbot 400	DQ229882	DQ229893
15	*Anzina carneonivea* (Anzi) Scheid.	Italy	SAG 26.81	AM900490	AM906042
15	*Lepraria neglecta* (Nyl.) Lettau	Ukraine	Bayerová 3600	AM906013	AM906044a
15	*Lepraria neglecta* (Nyl.) Lettau	Ukraine	Bayerová 3606	AM900941	AM906043a
16	*Cladonia fruticulosa* Kremp.	India	IH17	HE803051b	–
16	*Cladonia* cf. *bacillaris* (Ach.) Nyl.	USA, PA	Nelsen 3950	DQ229878	DQ229892
16	*Lepraria alpina* (de Lesd.) Tretiach & Baruffo	Czech Republic	Peksa 192	AM906010	AM906039
16	*Lepraria caesioalba* (de Lesd.) J.R.Laundon	Czech Republic	Peksa 194	AM906009	AM906038
16	*Lepraria caesioalba* (de Lesd.) J.R.Laundon	Czech Republic	Peksa 233	AM906006	AM906035
16	*Lepraria caesioalba* (de Lesd.) J.R.Laundon	Czech Republic	Peksa 166	AM906008	AM906037

ITS internal transcribed spacers

[a] Identical sequences omitted in the final alignment used for the Bayesian analysis

[b] Newly obtained sequences

was performed using heuristic searches with 100 random sequence addition replicates, tree bisection and reconnection (TBR) swapping, and random addition of sequences (the number limited to 10,000 for each replicate). The weight to the characters has been assigned using the rescaled consistency index, in a scale from 0 to 1,000. New weights were based on the mean of the fit values for each character over all of the trees in memory.

To map the biogeographic information onto the phylogenetic tree, we prepared a dataset of 319 ITS rDNA sequences (obtained in this study and acquired from GenBank database) with known biogeographic data. The distribution of *Asterochloris* in particular continents was finally shown for those clades containing at least ten sequences with known origin.

5   Results and Discussion

5.1   Diversity of Asterochloris photobionts

In total, 57 natural samples of various *Cladonia* species were collected from five different areas in India and Nepal. However, the amplification of ITS rDNA region was successful in only 20 of these samples (Table 4.1). Unsuccessful amplification of more than half of the samples might have been caused by their age and storage conditions (some *Cladonia* samples were more than 4 years old) or by the presence of nonspecific inhibitors. Usually, single photobiont has been detected in each lichen sample. However, in three cases we found two different *Asterochloris* genotypes in the single lichen thallus (samples IH2, IH8, and IH21).

All *Cladonia* samples were found to be associated with green algae belonging to the genus *Asterochloris*. The Bayesian analysis of the concatenated ITS rDNA and actin type I dataset led to the recognition of 20 lineages designated as clades 1–16 (according to Škaloud and Peksa 2010), clades A4, A9 (according to Peksa and Škaloud 2011), and two novel clades I1 and I2 (Fig. 4.2). The newly obtained photobiont sequences were inferred in six clades (I1, I2, 1, 9, 12 and 16). Two novel clades I1 and I2, exclusively formed by photobionts of Indian *Cladonia* lichens, were genetically considerably different from all other known *Asterochloris* lineages. Therefore, they very probably represent new, undescribed photobiont species. The clade I1 consisted of six photobiont sequences obtained from four *Cladonia* species (*C. rangiferina, C. furcata, C. pyxidata*, and *C. corymbescens*) collected in the Himalayas at relatively high altitude (2,300–3,700 m asl; Fig. 4.3). The clade I2 comprised only three photobiont sequences obtained from *Cladonia* lichens collected in both the Himalayas (samples I4 and IH26) and South India (sample IH23). All three lichen samples were also collected at high altitudes (2,607–3,250 m asl). Interestingly, all photobionts were found in *Cladonia furcata*, suggesting their specificity for this fungal partner.

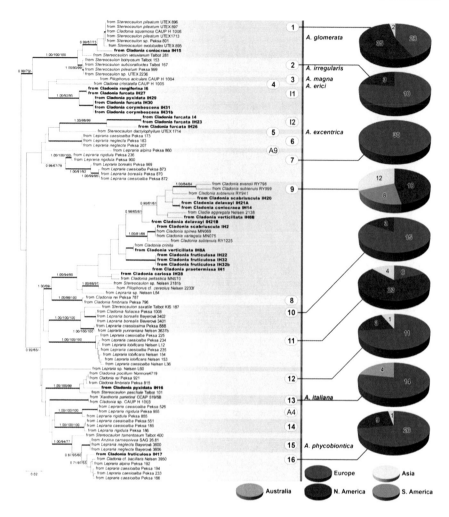

Fig. 4.2 Phylogenetic tree and biogeography of lichen photobiont *Asterochloris*. Bayesian analysis is based on the combined and partitioned internal transcribed spacer (ITS) ribosomal DNA (rDNA) and actin type I dataset using a HKY+I model for ITS1 and ITS2, F81 model for 5.8 ribosomal RNA (rRNA) partition, a HKY+Γ model for the actin-intron 206, GTR+Γ model for the actin intron 248 and K80+I model for the actin-exon partition. Values at the nodes indicate statistical support estimated by three methods: MrBayes posterior node probability (*left*), maximum likelihood bootstrap (in the *middle*) and maximum parsimony (*right*). *Thick branches* represent nodes receiving high Bayesian support (≥0.99) or consisting of genetically identical strains. New sequences from Indian *Cladonia* lichens are given in *bold*. Strain affiliation to 20 clades is indicated. Biogeography of selected lineages (those containing at least ten sequences with known origin) is shown next to the tree, including the total number of occurrences on each continent. Scale bar—estimated number of substitutions per site

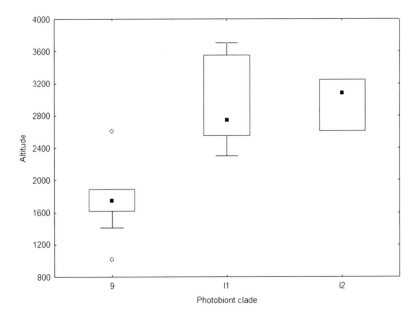

Fig. 4.3 Differences in the distribution of selected three *Asterochloris* clades along the altitudinal gradient. Box plots are based on altitudinal data of *Cladonia* samples analyzed in this study. All samples were collected in India and Nepal

The majority of investigated photobionts (found in 12 *Cladonia* samples belonging to 7 different species) were inferred in the clade 9. The clade is known as a lineage of North, Central, and South American lichen photobionts, having low specificity towards the lichen-forming fungi (it associates at least with 18 species from 5 lichen genera; Cordeiro et al. 2005; Nelsen and Gargas 2006; Piercey-Normore and DePriest 2001; Reis 2005; Yahr et al. 2004). Our lichen samples containing clade 9 photobionts were collected from various substrate types, such as bare soil, red hard soil, soil in coniferous forest, or rocks. In comparison with algal genotypes inferred in clades 11 and 12, clade 9 photobionts were found in the *Cladonia* samples collected at lower altitudes (1,014–2,607 m asl; Fig. 4.3). The remaining photobionts, found in *Cladonia* samples IH15, IH16, and IH 17 were inferred in three separate clades. The photobiont of *Cladonia coniocraea* (IH15) belongs to a very common species *Asterochloris glomerata* (clade 1). Two remaining photobionts, found in lichens *Cladonia pyxidata* (IH16) and *C. fruticulosa* (IH17) were inferred as members of clades 12 and 16, respectively.

5.2   Biogeography of Lichen Photobionts

During the last decade, biogeography of protists has become a highly controversial topic. It has been postulated that the small size, extremely large populations, and high dispersal potential of protists result in the cosmopolitan distribution of the vast majority of species (Finlay 2002; Finlay and Fenchel 2004). Conversely, the limited geographical distributions has been implied by Foissner (1999), based mainly on the observed restricted distribution of "flagship" species, i.e., species with easily recognizable morphologies whose presence/absence can be easily demonstrated (Foissner 2006, 2008). However, all protistan biogeographic studies have been based on the investigation of the free-living organisms.

Our study could bring valuable information about the distribution patterns of symbiotic protists. So far, the investigations on *Asterochloris* photobionts were predominantly conducted on European and American lichen samples, only a few data have been obtained from other continents (see Introduction). Therefore, addition of more than 20 newly generated *Asterochloris* sequences obtained from Indian *Cladonia* samples could improve the dataset for subsequent estimation of biogeographical patterns in lichen photobionts.

The biogeography of particular lichen photobiont clades is illustrated in Fig. 4.2. Only those clades containing at least ten sequences with known origin were analyzed. In general, the majority of clades show wide (eurychoric) distribution, i.e., they were found in two or three continents. For example, *Asterochloris glomerata*, the commonest species of the genus, display almost ubiquitous distribution. According to all published data so far, this species has been found in a number of various lichen taxa (almost 50 species from genera *Cladia, Cladonia, Stereocaulon, Pycnothelia, Diploschistes, Hertelidea*) collected in many different places in Europe, North America, and Asia. It has obviously wide ecological amplitude, occurring in lichens growing on a variety of different substrates and in various microclimatic conditions. Nevertheless, all records of *A. glomerata* originate from warm-temperate to (sub)arctic zones of northern hemisphere (similar to the clades 2, 10, 11, 12, and 16).

In comparison to other photobiont lineages, the clade 9 has extraordinary distribution pattern because of its absence in Europe (see Fig. 4.2). It is widely dispersed, reported from South to North America and Asia, however, all records occurred between latitudes 25°S (Brazil, Paraná) and 36°N (USA, North Carolina). Thus, the algae from clade 9 probably prefer tropical to warm-temperate climate. This fact could explain their absence in European samples (only warm Mediterranean regions of Europe can comply with such criterion, however, they have been poorly investigated for *Asterochloris* photobionts so far).

The earlier mentioned *Asterochloris* lineages exhibit wide distribution; nevertheless, their habitat area seems to be more or less restricted. Our current data, together with the results of Fernandez-Mendoza et al. (2011), Helms (2003), Kroken and Taylor (2000), Muggia et al. (2008), and Peksa and Škaloud (2011) suggest that one of the most important factors influencing the distribution of eukaryotic photobionts is climate. Such climatic preferences influence the type and size of species habitat. There are reports on lineages of *Trebouxia* photobionts occurring predominantly in tropical regions (Helms 2003), on the other hand, other clades (haplotypes) exhibit polar (bipolar) distribution pattern (Fernandez-Mendoza et al. 2011).

Thus, it is obvious that at least some clades occur only in specific biomes or latitudes in general, across different continents. It is a question whether there is any photobiont lineage living in one continent or region only (endemic species). According to our data, three clades seem to have rather restricted distribution. Photobionts of clade 7 (30 samples) have been reported so far only from Europe. Similarly, the clades 11 and 12 seem to be restricted to Asia (India). According to Foissner (2006), the restricted distribution of protist species could be caused by either historic, biological, climatic, or habitat factors. The biogeography of clades 7, 11 and 12 cannot be affected by the limited dispersal of their fungal partners. *Lepraria caesioalba* and *L. rigidula* (mycobionts of clade 7 algae), as well as *Cladonia furcata*, *C. rangiferina* and *C. pyxidata* (mycobionts of 11 and 12 algae) represent lichens with very wide to cosmopolitan distribution (Smith et al. 2009) and many of them disperse intensively via vegetative propagules (soredia, granules) which provide a possibility of intensive dispersal of both mycobionts and photobionts. Moreover, we cannot rule out the simple dispersal of photobionts independent of a fungus. *Asterochloris*, a unicellular green alga, asexually reproducing by high number of aplanospores (Škaloud and Peksa 2010) has virtually unlimited dispersal capacity. It is well supported by its common distribution and ubiquity of the majority of its species. Therefore, the restricted distribution of photobiont clades 7, 11, and 12 cannot be explained by either historic or biological factors. More likely, the clades are restricted in their distribution by having specific climatic or habitat preferences. The clade 7 photobionts, so far reported only from Europe, have been recently demonstrated to be significantly associated with ombrophobic lichens (i.e., growing in fully rain-sheltered sites, where the vapour is the only available source of water) growing predominantly on the bark of broadleaf trees in temperate belt. It is highly probable that further investigation of photobiont diversity in bark-associated greenalgal lichens conducted in other continents than Europe would reveal much wider distribution of this clade.

6 Conclusion

This study revealed significant photobiont diversity in *Cladonia* lichens collected in India and Nepal. The discovery of two novel, not yet reported clades emphasizes the large hidden diversity of lichen photobionts. Despite the fact that we investigated symbiotic organisms, almost all *Asterochloris* lineages exhibit eurychoric distribution. We suppose that the existence of several *Asterochloris* clades so far reported from single continent is affected by limited sampling and specific climatic or habitat preferences rather than by restricted distribution patterns. It is increasingly evident that the distinct preferences for environmental factors, not the dispersal barriers, shape the global distribution patterns of lichen photobionts. Consequently, narrow ecological preferences of lichen photobionts could to a certain extent determine the distribution pattern of the entire lichen association.

References

Aoki M, Nakano T, Kanda H, Deguchi H (1998) Photobionts isolated from Antarctic lichens. J Marine Biotechnol 6:39–43

Bačkor M, Peksa O, Škaloud P, Bačkorová M (2010) Photobiont diversity in lichens from metal-rich substrata based on ITS rDNA sequences. Ecotoxicol Environ Saf 73:603–612

Bailey RH (1976) Ecological aspects of dispersal and establishment in lichens. In: Brown DH, Bailey RH, Hawksworth DL (eds) Lichenology: Progress and Problems, Academic Press, London

Beck A (1999) Photobiont inventory of a lichen community growing on heavy-metal-rich rock. Lichenologist 31:501–510

Blaha J, Baloch E, Grube M (2006) High photobiont diversity in symbioses of the euryoecious lichen *Lecanora rupicola* (Lecanoraceae, Ascomycota). Biological J Linn Soc 88:283–293

Büdel B, Scheidegger C (2008) Thallus morphology and anatomy. In: Nash TH (ed) Lichen Biology. Cambridge University Press, Cambridge

Cordeiro LMC, Reis RA, Cruz LM, Stocker-Wörgötter E, Grube M, Iacomini M (2005) Molecular studies of photobionts of selected lichens from the coastal vegetation of Brazil. FEMS Microbiol Ecol 54:381–390

DePriest PT (2004) Early molecular investigations of lichen-forming symbionts: 1986–2001. Ann Rev Microbiol 58:273–301

Doering M, Piercey-Normore MD (2009) Genetically divergent algae shape an epiphytic lichen community on Jack Pine in Manitoba. Lichenologist 41:69–80

Doyle JJ, Doyle JL (1987) A rapid DNA isolation procedure for small quantities of fresh leaf tissue. Phytochem Bull 19:11–15

Engelen A, Convey P, Ott S (2010) Life history strategy of *Lepraria borealis* at an Antarctic inland site, Coal Nunatak. Lichenologist 42:339–346

Fernandez-Mendoza F, Domaschke S, Garciá MA, Jordan P, Martín MP, Printzen C (2011) Population structure of mycobionts and photobionts of the widespread lichen *Cetraria aculeata*. Mol Ecol 20:1208–1232

Finlay BJ (2002) Global dispersal of free-living microbial eukaryote species. Science 296:1061–1063

Finlay BJ, Clarke KJ (1999) Ubiquitous dispersal of microbial species. Nature 400:828

Finlay BJ, Fenchel T (2004) The Ubiquity of Small Species: Patterns of Local and Global Diversity. Bioscience 54:777–784

Foissner W (1999) Protist diversity: estimates of the near-imponderable. Protist 150:363–368

Foissner W (2006) Biogeography and dispersal of micro-organisms: a review emphasizing protests. Acta Protozool 45:111–136

Foissner W (2008) Protist diversity and distribution: some basic consideratons. Biodivers Conserv 17:235–142

Galloway DJ (2008). Lichen biogeography. In: Nash TH (ed) Lichen Biology. Cambridge University Press, Cambridge

Guzow-Krzemińska B (2006) Photobiont flexibility in the lichen *Protoparmeliopsis muralis* as revealed by ITS rDNA analyses. Lichenologist 38:469–476

Hauck M, Helms G, Friedl T (2007) Photobiont selectivity in the epiphytic lichens *Hypogymnia physodes* and *Lecanora conizaeoides*. Lichenologist 39:195–204

Helms G (2003) Taxonomy and symbiosis in association of *Physaceae* and *Trebouxia*. Dissertation, University of Gottingen, Germany

Hepperle D (2004) SeqAssem©. A sequence analysis tool, contig assembler and trace data visualization tool for molecular sequences. http://www.sequentix.de

Kroken S, Taylor JW (2000) Phylogenetic species, reproductive mode, and specificity of the green alga *Trebouxia* forming lichens with fungal genus *Letharia*. Bryologist 103:645–660

Muggia L, Grube M, Tretiach M (2008) Genetic diversity and photobiont associations in selected taxa of the *Tephromela* atra group (Lecanorales, lichenised Ascomycota). Mycol Prog 7:147–160.

Nelsen MP, Gargas A (2006) Actin type intron sequences increase phylogenetic resolution: An example from *Asterochloris* (Chlorophyta: Trebouxiphceae). Lichenologist 38:35–440

Nelsen MP, Gargas A (2008) Dissociation and horizontal transmission of co-dispersed lichen symbionts in the genus *Lepraria* (Lecanorales: Stereocaulaceae). New Phytol 177:264–275

Nelsen MP, Gargas A (2009) Symbiont flexibility in *Thamnolia vermicularis* (Pertusariales, Icmadophilaceae). Bryologist 112:404–417

Nyati S (2007) Photobiont diversity in *Teloschistaceae* (Lecanoromycetes). Dissertation, Universität Zürich, Switzerland

O'Brien HE, Miadlikowska J, Lutzoni F (2005) Assessing host specialization in symbiotic cyanobacteria associated with four closely related species of the lichen fungus Peltigera. Eur J Phycol 40:363–378

Otálora MAG, Martínez I, O'Brien H, Molina MC, Aragón G, Lutzoni F (2010) Multiple origins of high reciprocal symbiotic specificity at an intercontinental spatial scale among gelatinous lichens (*Collemataceae*, Lecanoromycetes). Mol Phylogenet Evol 56:1089–1095

Paulsrud P, Rikkinen J, Lindblad P (2000) Spatial patterns of photobiont diversity in some Nostoc-containing lichens. New Phytol 146:291–299

Peksa O, Škaloud P (2011) Do photobionts influence the ecology of lichens? A case study of environmental preferences in symbiotic green alga *Asterochloris* (*Trebouxiophyceae*). Mol Ecol 20:3936–3948

Piercey-Normore MD (2004) Selection of algal genotypes by three species of lichen fungi in the genus *Cladonia*. Can J Botany 82:947–961

Piercey-Normore MD (2006) The lichen-forming ascomycete *Evernia mesomorpha* associates with multiple genotypes of *Trebouxia jamesii*. New Phytol 169:331–344

Piercey-Normore MD (2009) Vegetatively reproducing fungi in three genera of the *Parmeliaceae* share divergent algal partners. Bryologist 112:773–785

Piercey-Normore MD, DePriest PT (2001) Algal-switching among lichen symbioses. Am J Bot 88:1490–1498

Reis RA (2005) Estudo filogenético de fotobiontes de liquens; isolamento e cultivo de simbiontes liquênicos; estudo comparativo de polissacarídeos e ácidos graxos do liquens Teloschistes flavicans e seus simbiontes. Tese de Doutorado apresentada, Universidade Federal do Paraná. Ph. D. thesis.

Rikkinen J, Oksanen I, Lohtander K (2002) Lichen guilds share related cyanobacterial symbionts. Science 297:357

Romeike J, Friedl T, Helms G, Ott S (2002) Genetic diversity of algal and fungal partners in four species of *Umbilicaria* (Lichenized Ascomycetes) along a transect of the Antarctic Peninsula. Mol Biol Evol 19:1209–1217

Ronquist F, Huelsenbeck JP (2003) MrBayes 3: Bayesian phylogenetic inference under mixed models. Bioinformatics 19:1572–1574

Škaloud P, Peksa O (2010) Evolutionary inferences based on ITS rDNA and actin sequences reveal extensive diversity of the common lichen alga *Asterochloris* (Trebouxiophyceae, Chlorophyta). Mol Phyl Evol 54:36–46

Smith CW, Aptroot A, Coppins BJ, Fletcher A, Gilbert OL, James PW, Wolseley PA (2009) The lichens of Great Britain and Ireland. British Lichen Society, London

Swofford DL (2002) In: PAUP*. Phylogenetics analyses using parsimony (and other methods). Version 4.Sinauer Associates, Sunderland, Massachusetts

White TJ, Bruns T, Lee S, Taylor J (1990) Amplification and direct sequencing of fungal ribosomal RNA genes for phylogenetics. In: Innis MA, Gelfand DH, Sninsky JJ, White TJ (eds) A Guide to Methods and Applications. Academic Press, San Diego

Wirtz N, Lumbsch HT, Green TGA, Türk R, Pintado A, Sancho L, Schroeter B (2003) Lichen fungi have low cyanobiont selectivity in maritime Antarctica. New Phytol 160:177–183

Wornik S, Grube M (2010) Joint dispersal does dot imply maintenance of partnerships in lichen symbioses. Microbial Ecol 59:150–157

Yahr R, Vilgalys R, DePriest PT (2004) Strong fungal specificity and selectivity for algal symbionts in Florida scrub *Cladonia* lichens. Mol Ecol 13:3367–3378

Zwickl DJ (2006) Genetic algorithm approaches for the phylogenetic analyses of large biological sequence datasets under the maximum likelihood criterion. Dissertation, University of Texas

Chapter 5
Photobiont Diversity of Soil Crust Lichens Along Substrate Ecology and Altitudinal Gradients in Himalayas: A Case Study from Garhwal Himalaya

Voytsekhovich Anna, Dymytrova Lyudmyla, Himanshu Rai and Dalip Kumar Upreti

1 Introduction

Lichens are considered as the most studied example of coevolution of two (or more) organisms. They are a symbiotic association of a single fungal species (mycobiont) and one or several species of green algae (chlorobiont) or cyanobacteria (cyanobiont), or sometimes even both green alga and cyanobacterium. The group of lichen-forming fungi consists of about 20,000 species (Kirk et al. 2008), whereas the number of the photobionts is only about 156 species from 56 genera (Voytsekhovich 2013). Most photobionts belong to green algae (Chlorophyta—116 species, 73.9 % of all photobiont diversity) and cyanobacteria (Cyanoprokaryota—35 species, 22.3 %).

Because of its algal component, the heterotrophic lichen-forming fungus becomes an autotrophic association, which requires for its existence only water, air, minerals and substrate. Such association and its further coevolution allowed lichens to occupy the most unfavourable habitats as bare rocks and deserts, and become a thriving group with high taxonomic diversity.

The Garhwal Himalaya harbours numerous habitats with a wide environmental heterogeneity in terms of light intensity, precipitation, humidity and temperature. The varied climatic conditions of the region support lichens of different growth forms. Among the various habitat subsets of lichens in Garhwal Himalayas, soil-inhabiting terricolous lichens have been found to be appropriate indicators of habitat heterogeneity and zooanthropogenic pressures (Rai et al. 2011, 2012). Although

V. Anna (✉) · D. Lyudmyla
Department of Lichenology and Bryology, M.H. Kholodny Institute of Botany,
National Academy of Sciences of Ukraine, 2 Tereshchenkivska st., 01601, Kyiv, Ukraine
e-mail: trebouxia@gmail.com

H. Rai · D. K. Upreti
Lichenology laboratory, Plant Diversity, Systematics and Herbarium Division
CSIR-National Botanical Research Institute,
Rana Pratap Marg, Lucknow
Uttar Pradesh-226001, INDIA

H. Rai, D. K. Upreti (eds.), *Terricolous Lichens in India,*
DOI 10.1007/978-1-4614-8736-4_5, © Springer Science+Business Media New York 2014

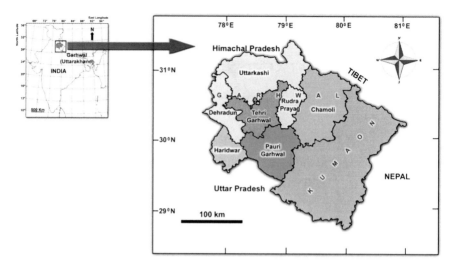

Fig. 5.1 Location map of Garhwal and its constituent districts in state of Uttarakhand (maps and their grid over lay are based on the 1:1,000,000 state map of Uttarakhand, first edition, Survey of India, Department of Science and Technology, Government of India, 2001)

there have been few studies dealing with the taxonomic and biomonitoring aspects of terricolous lichens of the region, no study is yet done on the photobiont diversity, their role in overall ecological preferences of terricolous lichens. Therefore, the present study aims (1) to analyze the lichen photobiont diversity of terricolous lichens of Garhwal Himalaya at the generic level; (2) to determine whether the photobiont depends on the altitude above sea level (asl) at which the lichen grows; and (3) to study and compare the photobiont composition of different ecological subgroups of terricolous lichens.

2 Materials and Methods

2.1 Study Area

Garhwal is a region and administrative division of Uttarakhand state, India, lying in Himalaya, and is bounded on the north by Tibet, on the east by Kumaon region, on the south by Uttar Pradesh state and on the west by Himachal Pradesh state (Fig. 5.1). Garhwal region comprises seven districts i.e., Chamoli, Dehradun, Haridwar, Pauri Garhwal, Rudraprayag, Tehri Garhwal, and Uttarkashi (Fig. 5.1). The administrative centre for Garhwal division is the town of Pauri.

The region consists almost entirely of rugged mountain ranges running in all directions, and separated by narrow valleys which, in some cases, become deep gorges or ravines. The highest mountains are in the north, the principal peaks being Nanda Devi (7,821 m), Kamet (7,746 m), Trisul (7,127 m), Badrinath (7,074 m), Dunagiri

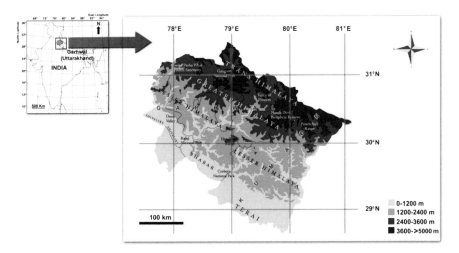

Fig. 5.2 Simple relief map of Uttarakhand showing altitudinal variations of Garhwal Himalaya (modified after Rawat 2010, grid over lay are based on the 1:1,000,000 state map of Uttarakhand, first edition, Survey of India, Department of Science and Technology, Government of India, 2001)

(7,066 m) and Kedarnath (6,966 m). The Himalaya in this region is represented by the outer Himalayas/Shiwalik Range (500–1,200 m) in the district of Dehradun, Haridwar and the southern area of district Pauri; the Lesser or Middle Himalayas (3,700–4,500 m) in the districts of Uttarkashi, northern Pauri, Tehri, Rudraprayag and Chamoli; and the Greater Himalaya or the inner Himalaya (above 4,500 m) in the districts of Uttarkashi, Rudraprayag and Chamoli (Fig. 5.2) (Singh and Singh 1987).

Climate of Garhwal Himalaya ranges from subtropical (i.e. Haridwar) to temperate (i.e. Dehradun) in foothills to temperate-alpine in higher elevations (i.e. Tehri, Pauri, Rudraprayag, Uttarkashi, Chamoli). Precipitation occurs mainly because of monsoons, in June-August. Average annual precipitation of the region is 1,550 mm (IMD 1989), which ranges from 600 to 2,350 mm (Sharma et al. 2010; Vishwakarma et al. 2012). Although there is daily orographic precipitation at higher altitudes (>2,700 m), maximum rainfall is received in the months of June to September (Khare et al. 2010). Average annual temperature ranges from 19-37 °C (Negi 2000; Rai et al. 2012). Minimum temperature easily dips to subzero levels at higher altitudes (up to − 19 °C) during November-February (Negi 2000; Rai et al. 2012). Higher altitudes receive their maximum snowfall during the months November to April and snowmelt is the major source of soil water prior to monsoons (Nautiyal et al. 2001; Rai et al. 2012).

Soils of the region are generally acidic (pH 4–5), coarse textured loam to sandy loam at lower altitudes to sandy at higher elevation (Singh and Singh 1987; Sundriyal 1992; Rai et al. 2012). The nutrient contents of the soil in terms of total organic carbon (TOC) and nitrogen (N) is poor and ranges from 0.7 to 4.0 % for TOC and 0.02–0.3 % for N (Singh and Singh 1987). The value of TOC and N is also influenced by local factors (microtopography, rate of litter degradation and input by anthropogenic sources) (Singh and Singh 1987).

Garhwal Himalaya is the constituent component of the Central Himalayan botanical region (Singh and Singh 1987). The vegetation in Garhwal Himalaya shows a succession of vegetation from tropical to the alpine. At foothills (1,000–2,000 m), the vegetation is deciduous type and main trees are oaks (*Quercus* spp.) and sal (*Shorea* spp.); between the altitude of 2,000 m and 3,000 m, the dominant vegetation is of moist temperate forests and main trees are oaks, pines (*Pinus* spp.), spruce (*Picea* spp.) and deodar (*Cedrus* spp.); and above 3,000 m alt, there is transition to alpine forests and pastures characterized by shrubs of *Rhododendron, Anthopogon* and *Juniperus* and herb species of *Anemone, Potentilla, Aster, Geranium, Meconopsis, Primula* and *Polemonium* (Rai et al. 2012).

Garhwal Himalaya harbours a rich diversity of lichens both in the terms of species and growth forms which occupy all available relevés of a habitat (Upreti 1998; Upreti and Negi 1998). Terricolous lichens constitute about 9% of total lichen species recorded from India and their distribution ranges from temperate (1,500–3,000 m) to alpine (>3,000 m) (Rai et al. 2011, 2012). In Garhwal Himalaya, soil lichens, however, constitute about 1.2% of total lichen biota (Negi 2000).

2.2   Data Sources

One hundred and fifty taxa of terricolous lichens, collected from Garhwal Himalaya and already lodged as voucher specimens (872 in total) in lichenology herbarium of CSIR-National Botanical Research Institute, Lucknow, Uttar Pradesh (India), were examined for the study. The photobiont diversity was analyzed using relevant literature available (Table 5.1).

2.3   Data Analysis

The data accumulated was analyzed using *univariate descriptive statistics* (mean, standard deviation, minimum and maximum) and results were summarized using box plots and stacked column bar plots.

3   Results and Discussion

3.1   Photobiont Composition of Terricolous Lichens in Garhwal Himalayas

The photobiont composition of investigated lichens consists of Chlorophyta (105 lichen taxa; 70% of investigated lichen specimens) and Cyanoprokaryota (45 taxa; 30%). Most of studied lichen-forming fungi were associated with *Asterochloris*

Table 5.1 Analyzed lichen genera and their photobionts according to the literature data

Lichen genus	Primary photobiont genus	Reference
Acarospora	*Trebouxia*	Beck (1999, 2002)
Allocetraria	*Trebouxia*	Ahmadjian (1993)
Bryoria	*Trebouxia*	Tarhanen et al. (1997)
Bulbothrix	*Trebouxia*	Hale (1974)
Cetraria	*Trebouxia*	Frenández-Mendoza et al. (2011)
Cetrelia	*Trebouxia*	Helms (2003)
Cladia	*Asterochloris*	Ahmadjian (1993)
Cladonia	*Asterochloris*	Piercey-Normore and De Priest (2001), Skaloud and Peksa (2010)
Coccocarpia	*Scytonema*	Santesson (1952)
Collema	*Nostoc*	Beck, Kasalicky and Rambold (2002), Otálora et al. (2010)
Dermatocarpon	*Diplosphaera*	Zeitler (1954), Thüs et al. (2011)
Endocarpon	*Diplosphaera*	Zeitler (1954), Thüs et al. (2011)
Evernia	*Asterochloris*	Ahmadjian (1993)
Everniastrum	*Trebouxia*	Nash et al. (2002)
Flavocetraria	*Trebouxia*	Opanowich and Grube (2004), Bačkor et al. (2010)
Flavoparmelia	*Trebouxia*	Ahmadjian (1993), Friedl et al. (2000)
Fuscopannaria	*Nostoc*	Awasthi (2007)
Heterodermia	*Trebouxia*	Moberg and Nash (1999)
Hypogymnia	*Trebouxia*	Romeike et al. (2002)
Hypotrachyna	*Trebouxia*	Ahmadjian (1993)
Lempholemma	*Nostoc*	Awasthi (2007)
Lepraria	*Asterochloris*	Skaloud and Peksa (2010), Engelen et al. (2010)
Leptogium	*Nostoc*	Otálora et al. (2010)
Melanelia	*Trebouxia*	Ahmadjian (1993)
Melanelixia	*Trebouxia*	Ahmadjian (1993)
Mycobilimbia	Chlorococcoid green alga	Ekman (2004), Kantvilas et al. (2005)
Nephroma	*Nostoc/Coccomyxa*	Lohtander et al. (2002), O'Brien et al. (2005), Otálora et al. (2010)
Parmelia	*Trebouxia*	Archibald (1975), Ahmadjian (1993)
Parmelinella	*Trebouxia*	Awasthi (2007)
Peccania	*Gloeocapsa*	Geitler (1933)
Peltigera	*Nostoc*	O'Brien et al. (2005)
Phaeophyscia	*Trebouxia*	Dahlkild et al. (2001), Helms et al. (2001), Romeike et al. (2002)
Physcia	*Trebouxia*	Dahlkild et al. (2001), Helms et al. (2001)
Physconia	*Trebouxia*	Dahlkild et al. (2001), Helms et al. (2001)
Ramalina	*Trebouxia*	Cordeiro et al. (2005)
Rhizoplaca	*Trebouxia*	Hildreth and Ahmadjian (1981)
Stereocaulon	*Asterochloris*	Duvigneaud (1955), Archibald (1975), Skaloud and Peksa (2010)

Table 5.1 (continued)

Lichen genus	Primary photobiont genus	Reference
Sticta	*Nostoc*	Otálora et al. (2010)
Thamnolia	*Trebouxia*	Ihda and Nakano (1995)
Toninia	*Trebouxia*	Ahmadjian (1993), Beck et al. (2002)
Umbilicaria	*Trebouxia*	Beck (1999), Engelen et al. (2010)
Xanthoparmelia	*Trebouxia*	Ahmadjian (1993)

Tscherm.-Woess (46 taxa, 436 specimens), *Trebouxia* Puym. (55 taxa, 113 specimens) and *Nostoc* Vaucher ex Bornet et Flahault (41 taxa, 302 specimens). A few lichens contain *Scytonema* C. Agardh ex Bornet et Flahault (3 lichen taxa, 15 specimens), *Diplosphaera* Bial. (2 taxa, 3 specimens) and *Gloeocapsa* Kütz. (1 taxon, 1 specimen) (Table 5.2). For two remaining lichens *Mycobilimbia hunana* and *M. philippina*, there is no accurate data on their photobiont composition in the literature. In some publications (e.g. Ekman 2004; Kantvilas et al. 2005), however, it is specified that these lichen-forming fungi are associated with chlorococcoid green alga.

Some lichen-forming fungi of the genera *Nephroma* Ach., *Peltigera* Willd., *Stereocaulon* Hoffm. and *Sticta* (Schreb.) Ach. have cephalodia and form photosymbiodemes, when the same thallus contains cyanobacterium and green alga simultaneously (Renner and Galloway 1982). For instance, *Nephroma* contains *Coccomyxa* sp. in its thallus and *Nostoc* sp. in internal cephalodia (Wetmore 1960; Lohtander et al. 2002). However, there is also evidence that the cephalodial content is quite variable, e.g. besides the primary photobiont from the genus *Asterochloris*, the lichen *Stereocaulon alpinum* Laurer contains cyanobacterium *Stigonema* sp. (Lamb 1951) in its cephalodia, whereas another species, *Stereocaulon pomiferum* P.A. Duvign., contains cyanobacterium *Anabaena* sp. (Duvigneaud 1955). The photobiont composition of *Stereocaulon* cannot be considered constant, because, according to Lamb (1951) and Duvigneaud (1955), its cephalodia contain different cyanobacteria, such as previously mentioned *Anabaena* Bory ex Bornet et Flahault, as well as *Aphanocapsa* Nägeli, *Gloeocapsa* Kütz., *Stigonema* Agardh ex Bornet et Flahault, *Calothrix* Agardh ex Bornet et Flahault, *Nostoc*, etc.

In contrast to photosymbiodemous lichens, *Lepraria* spp. occasionally form their thalli with several different species of green alga simultaneously, by including the additional photobionts directly into the thallus. Aoki et al. (1998) showed that the same thallus of *Lepraria* sp. contained *Trebouxia* cf. *impressa* Ahmadjian and *Elliptochloris bilobata* Tscherm.-Woess simultaneously. According to our data (Voytsekhovich et al. in press), one specimen of *Lepraria membranacea* (Dicks.) Vain. contained *Trebouxia incrustata* Ahmadjian ex Gärtner and *Chloroidium ellipsoideum* (Gerneck) Darienko et al. when other specimen was associated with *Trebouxia* sp. and *Stichococcus bacillaris* Nägeli. Due to an inconstant composition of additional photobionts, we consider only their primary photobiont.

Table 5.2 Photobiont diversity and distribution among the specimens, taxa and ecological subgroups of investigated terricolous lichens of Garhwal Himalaya, with reference to altitudinal gradient

Photobiont	Number of lichen specimens	Number of lichen taxa	Range of altitudes (m)	Ecological subgroup of terricolous lichens (Scheidegger and Clerc 2002)				
				Detriticolous–terricolous	Muscicolous–rupicolous	Muscicolous–terricolous	Terricolous	Terricolous–rupicolous
Asterochloris	436	46	560–4,200	9	30	18	233	146
Diplosphaera	3	2	3,000–3,871	0	0	0	1	2
Gloeocapsa	1	1	3,505	0	0	0	1	0
Green alga	2	2	1,350–2,438	0	0	0	0	2
Nostoc	302	41	1,000–4,500	0	32	18	139	113
Scytonema	15	3	1,200–4,115	0	1	0	2	12
Trebouxia	113	55	1,280–4,572	4	5	2	55	47
Total	872	150	–	13	68	38	431	322

Most of above-mentioned photobionts (*Diplosphaera, Gloeocapsa, Scytonema* and *Nostoc*) are well known as free-living terrestrial algae which usually can be found on the different substarata, i.e. rocks, tree bark, soil or even the lichen thalli (Hoffmann 1989; Ettl and Gärtner 1995; Nienow 1996; Kumar et al. 2011). These algae are known as facultative (optional) photobionts of lichens. In contrast, *Asterochloris* and *Trebouxia* can appear in lichen thalli or in their vicinity predominantly and belong to obligate lichen photobionts. Indeed, most species of these two genera occur only as lichen photobionts (Ettl and Gärtner 1995), although some of them, e.g. *Trebouxia corticola* (P. Archibald) Gärtner (Archibald 1975), *T. arboricola* Puym. (Bubrick et al. 1984), *Asterochloris irregularis* (Hildreth et Ahmadjian) Skaloud et Peksa (Mikhailyuk and Darienko 2011), *A. excentrica* (P. A. Archibald) Skaloud et Peksa (Voytsekhovich et al. in press), were occasionally observed in free-living state. Despite this, the validity of such "free-living" foundings is questionable owing to certain methodological problems, because the enrichment cultures of algae may be contaminated with lichen isidia and soredia.

3.2 Photobiont Distribution Within Different Ecological Subgroups of Terricolous Lichens

It is well known that species composition of cryptogamous communities (lichens, mosses and terrestrial algae) depends on physicochemical properties of the substrate (Barkman 1958; Oxner 1974; John 1989; Uher et al. 2005; Macedo et al. 2009). Therefore, some lichen species could be found only on one type of substrate. We compared and analyzed the photobionts of 872 specimens of terricolous lichens collected from the different substrate types: gravel, rocks, soil, mosses, decaying wood and plant debris, etc. For this purpose, we used the classification of Scheidegger and Clerc (2002), who divided terricolous lichens on five ecological subgroups according to the type of substrate on which they grow: (1) terricolous (Tr), species that grow on the ground directly on soil, sand, peat or humus; (2) muscicolous–terricolous (Mt), species growing on mosses which are rooted on soil or sand; (3) terricol–rupicole (Trp), species that grow on accumulated soil in rock crevices or rough surfaces of rock; (4) muscicolous–rupicolous (Mr), species that grow on mosses which are rooted directly on rock and have some accumulated soil or organic debris; (5) detriticolous–terricolous (Dt), growing directly on plant remains on the ground.

The largest number of studied lichen specimens belongs to terricolous (431 specimens) and terricolous–rupicolous (322 specimens) subgroups, whereas other subgroups are less numerous: muscicolous–rupicolous (68 specimens), muscicolous–terricolous (38 specimens) and detriticolous–terricolous (13 specimens).

Comparative analysis of lichen photobionts of different ecological subgroups of terricolous lichens in Garhwal Himalaya shows that the highest photobiont diversity belongs to terricolous and terricolous–rupicolous lichens, and the lowest to detriticolous–terricolous (Fig. 5.3). Generally, all subgroups are characterized by

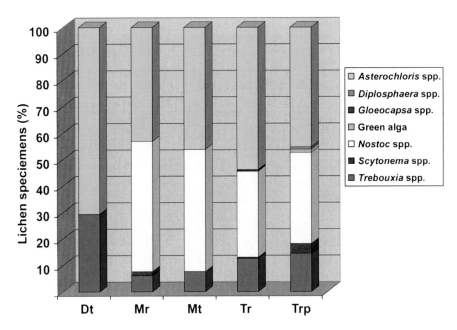

Fig. 5.3 Photobiont composition within different ecological subgroups in Garhwal Hymalayas. *Tr* terricolous, *Mt* muscicolous–terricolous, *Trp* terricol–rupicole, *Mr* muscicolous–rupicolous, *Dt* detriticolous–terricolous

a relatively high number of specimens associated with *Asterochloris*, although in muscicolous–rupicolous subgroup *Nostoc* is dominant. Remarkably, the photobiont composition of detriticolous–terricolous lichens consists of green algae entirely.

The obtained pattern of the photobionts to subgroups can be explained by the peculiarities of photobiont physiology. It is well known that cyanoprokariotic organisms and green algae require different conditions to start the process of photosynthesis. For instance, at high concentration of osmotically active substances in the cell sap and increased density of the cytoplasm, the cells of green photobiont can assimilate water even in the form of water vapour (Lange et al. 1998; Lange et al. 2006). Therefore, to resume photosynthesis after drying, the cells of green photobiont need high air humidity (Büdel and Lange 1991).

In contrast, to start the process of photosynthesis, the cyanobacterial cells require drip liquid water (Büdel and Lange 1991; Lange et al. 1998; Honegger 1991). However, this physiological "flaw" of cyanoprokaryotes is successfully compensated with their other features, such as the ability to store water in the mucilage (and thus extend the wet period) and the ability to photosynthesize at much higher temperature and light than eukaryotic algae (Lange et al. 1998). According to this, the lichens with green algal photobionts are more likely to prefer habitats with fog and constant water vapour, which could even be exposed areas of rocks, rocky soil or bark. Cyanolichens prefer habitats with periodic congestion of liquid water, such as cracks in the rocks, deepenings in rocks and soil as well as locations of water flow, although these habitats can also be sun-exposed.

Thereby, many lichen species of such ecological subgroups of terricolous lichens, such as terricolous, terricolous–rupicolous, muscicolous–rupicolous and muscicolous–terricolous that grow on the substrates which can hold water (permanently moisturized soil, deepenings in rocks with soil or clumps of mosses), are associated with cyanobionts. Lichens of detriticolous–terricolous subgroup that grow on decaying rests of different plants lying on the ground are associated with green algae because their substrate hardly can hold enough liquid water, which is so necessary for cyanobacteria.

4   The Photobiont Distribution of Terricolous Lichens Along Altitudinal Gradients in Garhwal Himalayas

To determine if the photobiont depends on the altitude at which the lichen grows, we analyzed the photobiont composition of 866 lichen specimens collected from Garhwal Himalayas. All analyzed specimens were collected within the altitude range 560–4,572 m (Table 5.2). Lichens associated with *Diplosphaera*, *Gloeocapsa* and chlorococcoid green algal photobiont were excluded from analysis due to insufficient number of their specimens.

The results of statistical analysis (Fig. 5.4) demonstrate that species richness of lichens with various photobionts is different at low and high altitudes. For instance, at 560–1,000 m asl, only lichens with *Asterochloris* were found; above 1,000 m, cyanolichens with *Nostoc* also occurred; and eventually, at the altitude of 1,200–1,300 m, lichens with *Scytonema* and *Trebouxia* gradually appeared. However, most of the lichen species with *Asterochloris* were found in the range 2,300–3,700 m, with *Scytonema* at 1,700–3,900 m, *Nostoc* at 2,100–3,500 m and *Trebouxia* at 2,800–4,000. Considering that the highest lichen richness was within the range of 2,800–3,500-m altitude, it is related to climatic factors, i.e. light intensity, humidity/precipitation and temperature.

It is known that different photobionts require different conditions for their growth. For instance, some *Trebouxia* species need the temperature 5–10 °C (Tschermak-Woess 1989; Voytsekhovich and Kashevarov 2010), other green photobionts prefer 18–21 °C (Ahmadjian 1993), whereas cyanobacteria require 24–30 °C (Kumar et al. 2011). However, some cyanoprokaryotes are also able to withstand extremely low temperatures. For instance, well-known cyanobiont *Nostoc commune* has an excellent tolerance to low temperatures and maintains high metabolism at − 18 to 20 °C; in desiccated state, it survives even −269 °C (Sand-Jensen and Jespersen 2012). The temperature gradient is consistent with the definition of altitude in the mountains (Loppi et al. 1997; Wang et al. 2011). In other words, the higher the altitude asl, the lower the air temperature. Light conditions also play an important role in lichen distribution (Jairus et al. 2009). Most lichens avoid both direct solar radiation

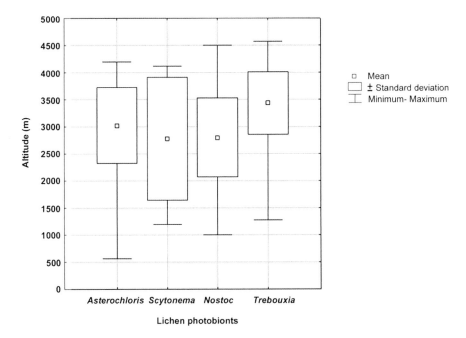

Fig. 5.4 Terricolous lichen photobiont distribution along altitude gradients in Garhwal Himalayas

(Barkman 1958) and excessive shade (Aude and Poulsen 2000; Moning et al. 2009). Therefore, the highest lichen diversity does not occur on the tops of mountains or foothills, but in a certain interval between these two points. Unfortunately, owing to the fact that we have not performed specific measurements of temperature and light conditions, our data are not sufficient to draw any serious conclusions about the effect of temperature and insolation on the photobiont composition of terricolous lichens from Garhwal Himalaya. Therefore, the issue remains relevant and requires further studies.

Because many lichens are aerohygrophytic species (Pirintsos et al. 1995; Scheidegger et al. 1995), high humidity from fogs and low-lying clouds at high altitude favour the occurrence of lichen species. However, as the annual precipitation and temperature decreases with elevation above 3,500–4,000 m, the growth of photosynthetic organisms at elevations above 4,000 m is seriously limited by water supply (Higuchi et al. 1982; Morales et al. 2004; Barros et al. 2006). Therefore, gradient of altitudes is a very important factor explaining lichen species richness and composition of their photobionts (Will-Wolf et al. 2006; Moning et al. 2009). Cyanobionts dominate those niches which can hold water for longer period, whereas green algal chlorobionts dominate the rest. Our results confirm previous studies, which pointed out the importance of climatic factors connected with the altitude gradient for the diversity of lichen communities (Barkman 1958; Pirintsos et al. 1995; Heylen et al. 2005; Werth et al. 2005).

References

Ahmadjian V (1993) The Lichen Symbiosis. Wiley, New York, 250 p

Aoki M, Nakano T, Kanda H, Deguchi H (1998) Photobionts isolated from Antarctic lichens. J Mar Biotechnol 6:39–43

Archibald PA (1975) Trebouxia de Puymaly (Chlorophyceae, Chlorococcales) and Pseudotrebouxia gen. nov. (Chlorophyceae, Chlorococcales). Phycologia 14:125–137

Aude E, Poulsen RS (2000) Influence of management on the species composition of epiphytic cryptogams in Danish Fagus forests. App Veg Sci 3:81–88

Awasthi DD (2007) A compendium of the macrolichens from India, Nepal and SriLanka. Bishen Singh Mahendra Pal Singh, Dehra Dun. p 580

Bačkor M, Peksa O, Škaloud P, Bačkorová M (2010) Photobiont diversity in lichens from metal-rich substrata based on ITS rDNA sequences. Ecotox Environ Safe 73:603–612

Barkman JJ (1958) Phytosociology and ecology of cryptogamic epiphytes. Van Gorcum, Assen. Netherlands, 628 p

Barros AP, Chiao S, Lang TJ, Burbank D, Putkonen J (2006) From weather to climate Seasonal and interannual variability of storms and implications for erosion processes in the Himalaya, In: Willett SD, Hovius N, Brandon MT, Fisher D (eds) Tectonics, Climate, and Landscape Evolution: Geological Society of America Special Paper 398: p 17–38

Beck A (1999) Photobiont inventory of a lichen community growing on heavy-metal-rich rock. Lichenologist 31:501–510

Beck A (2002) Selectivität der Symbionten schwermetalltoleranter Flechten.—Dissertation. München, 196 p, ISBN: 3-9808102-0–8

Beck A, Kasalicky T, Rambold G (2002) Myco-photobiontal selection in a Mediterranean cryptogam community with Fulgensia fulgida. New Phytol 153:317–326

Bubrick P, Galun M, Frensdorff A (1984) Observations on free-living Trebouxia de Puymaly and Pseudotrebouxia Archibald, and evidence that both symbionts from Xanthoria parietina (L.) Th. Fr. can be found free-living in nature. New Phytol 97:455–462

Büdel B, Lange OL (1991) Water status of green and blue-green phycobionts in lichen thalli after hydration by water vapor uptake: do they become turgid? Bot Acta 104:361–366

Cordeiro LMC, Reis RA, Cruz LM, Stocker-Worgotter E, Grube M, Iacomini M (2005) Molecular studies of photobionts of selected lichens from coastal vegetation of Brasil. FEMS Microbiol Ecol 54:381–390

Dahkild A, Källersjö M, Lohtander K, Tehler A (2001) Photobiont diversity in the Physciaceae (Lecanorales). Bryologist 104(4):527–536

Duvigneaud P (1955) Les Stereocaulon des hautes montanges du Kivu. Lejeunia Mem 14:1–9

Ekman S (2004) Mycobilimbia. In: Nash TH, Rayan BD, Gries C, Bungartz F (eds) Lichen flora of the great Sonoran Desert Region: most of the microlichens, balance of the macrolichens, and the lichenicolous fungi. Vol. 2. Arizona: Lichens Unlimited, Arizona State University, 365–367

Engelen A, Convey P, Ott S (2010) Life history strategy of Lepraria borealis at an Antarctic inland site, Coal Nunatak. Lichenologist 42:339–346

Ettl H, Gärtner G (1995) Syllabus der Boden-, Luft-, und Flechtenalgen. – Gustav Fischer, Stuttgart, Jena, New York, 710 p

Fernández-Mendoza F, Domaschke S, García MÁ, Jordan P, Martín M Printzen C (2011) Population structure of mycobionts and photobionts of the widespread lichen Cetraria aculeata. Mol Ecol 20:1208–1232

Friedl T, Besendahl A, Pfeiffer P, Bhattacharya D (2000) The distribution of group I introns in lichen algae suggests that lichenization facilitates intron lateral transfer. Mol Phylogenet Evol 14:342–352

Geitler L (1933) Beiträge zur Kenntnis der Flechtensymbiose I-III. Arch. Protistenk. 80:378–409

Hale ME (1974) Bulbothrix, Parmelina, Relicina and Xanthoparmelia, four new genera in the Parmeliaceae. Phytologia 28:479–490

Helms G (2003) Taxonomy and symbiosis in associations of Physciaceae and Trebouxia. Dissertation. University of Göttingen, Germany, 155 p

Helms G, Friedl T, Rambold G, Mayrhofer H (2001) Identification of photobionts from the lichen family Physciaceae using algal-specific ITS rDNA sequencing. Lichenologist 33:73–86

Heylen O, Hermy M, Schrevens E (2005) Determinants of cryptogamic epiphyte diversity in a river valley (Flanders). Biol Conserv 126:371–382

Higuchi K, Ageta Y, Yasunari T, Inoue J (1982) Characteristics of precipitation during the monsoon season in high-mountain areas of the Nepal Himalaya. Hydrological Aspects of Alpine and High Mountain Areas (Proceedings of the Exeter Symposium, July 1982). IAHS Publ. no. 138: 2130

Hildreth KC, Ahmadjian V (1981) A study Trebouxia and Pseudotrebouxia isolates from different lichens. Lichenologist 13:65–86

Hoffmann L (1989) Algae of terrestrial habitats. Botanical Review 55:77–105

Honegger R (1991) Functional aspects of the lichen symbiosis. Annu. Rev. Plant Mol Biol 42:553–578

Ihda T, Nakano T (1995) Temperature characteristics of photobionts isolated from alpine lichens. Proc. NIPR Symp. Polar Biol 8:205–206

IMD (Indian Meteorological Department) (1989) Climate of Uttar Pradesh, pp 372–375. Government of India Publication.

Jairus K, Lõhmus A, Lõhmus P (2009) Lichen acclimatization on retention trees: a conservation physiology lesson. J App Ecol 46:930–936

John EA (1989) An assessment of the role of biotic interactions and dynamic processes in the organization of species in a saxicolous lichen community. Can J Botany 67:2025–2037

Kantvilas G, Messuti MI, Lumbsch HT (2005) Additions to the genus Mycobilimbia s. lat. from the Southern Hemisphere. Lichenologist 37:251259

Khare R, Rai H, Upreti DK, Gupta RK (2010) Soil Lichens as indicator of trampling in high altitude grassland of Garhwal, Western Himalaya, India. Fourth National Conference on Plants and Environmental Pollution 8–11 Dec. 2010, p 135–136

Kirk PM, Cannon PF, Minter DW, Stalpers JA (2008) Dictionary of the Fungi.—Trowbridge: Cromwell Press, 771 p

Kumar M, Gupta RK, Bhatt AB, Tiwari SC (2011) Epiphytic cyanobacterial diversity in the sub-Himalayan belt of Garhwal region of Uttarakhand, India. Botanica Orientalis 8:77–89

Lamb IM (1951) On the morphology, phylogeny, and taxonomy of the lichen genus Stereocaulon. Can J Botany 29:522–536

Lange OL, Belnap J, Reichenberger H (1998) Photosynthesis of the cyanobacterial soil-crust lichen Collema tenax from arid lands in southern Utah, USA: role of water content on light and temperature responses of CO_2 exchange. Funct Ecol 12:195–202

Lange OL, Green A, Melzer B, Meyer A, Zellner H (2006) Water relations and CO_2 exchange of the terrestrial lichen Teloschistes capensis in the Namib fog desert: Measurements during two seasons in the field and under controlled conditions. Flora- Morphol Distrib Funct Ecol Plant 201:268–280

Lohtander K, Oksanen I, Rikkinen J (2002) A phylogenetic study of Nephroma (lichen-forming Ascomycota). Mycol Res 106:777–787

Loppi S, Pirintsos SA, Dominicis VD (1997) Analysis of the distribution of epiphytic lichens on Quercus pubescens along an altitudinal gradient in a mediterranean area (Tuscany, Central Italy). Israel J Plant Sci 45:53–58

Macedo MF, Miller AZ, Dionísio A, Saiz-Jimenez C (2009) Biodiversity of cyanobacteria and green algae on monuments in the Mediterranean Basin: an overview. Microbiology 155:3476–3490

Mikhailyuk TI, Darienko TM (2011) Algae of terrestrial habitats in National Nature Park Hutsulshchyna. In: Protected areas of Ukraine. Flora, vol 9. Phytosociotsenter, Kyiv, p 142–151

Moberg R, Nash TH III (1999) The genus Heterodermia in the Sonoran Desert area. Bryologist 102:1–14

Moning C, Werth S, Dziock F, Bässler C, Bradtka J, Hothorn T, Müller J (2009) Lichen diversity in temperate montane forests is influenced by forest structure more than climate. Forest Ecol Manag 258:745–751

Morales MS, Villalba R, Grau HR, Paolini L (2004) Rainfall-controlled tree growth in high-eleva-
    tions subtropical treelines. Ecology 85:3080–3089
Nash III TH, Ryan BD, Gries C, Bungartz F (eds) (2002) Lichen Flora of the Greater Sonoran
    Desert Region, Vol. 1. Tempe, Arizona: Lichens Unlimited, Arizona State University
Nautiyal MC, Nautiyal BP, Prakash V (2001) Phenology and growth form distribution in an alpine
    pasture at Tungnath, Garhwal, Himalaya. Mt Res Dev 21:168–174
Negi H.R. (2000) On the patterns of abundance and diversity of macrolichens of Chopta–Tungnath
    in Garhwal Himalaya. J Bioscie 25:367–378
Nienow JA (1996) Ecology of subaerial algae. Nova Hedwigia 112:537–552
O'Brien HE, Miadlikowska J, Lutzoni F (2005) Assessing host specialization in the symbiotic
    cyanobacteria associated with four closely related species of the lichen fungus Peltigera. Eur
    J Phycol 40:363–378
Opanowicz M, Grube M (2004) Photobiont genetic variation in Flavocetraria nivalis from Poland
    (Parmeliaceae, lichenized Ascomycota). Lichenologist 36:125–131
Otálora MAG, Martínez I, O'Brien H, Molina MC, Aragón G, Lutzoni F (2010) Multiple origins
    of high reciprocal symbiotic specificity at an intercontinental spatial scale among gelatinous
    lichens (Collemataceae, Lecanoromycetes). Mol Phylogenet Evol 56:1089–1095
Oxner AN (1974) Handbook of the lichens of the U.S.S.R. 2. Morphology, systematic and geogra-
    phical distribution. Nauka, Leningrad, 284 p (in Russian)
Piercey-Normore M, De Priest PT (2001) Algal switching among lichen symbioses. Am J Bot
    88:1490–1498
Pirintsos SA, Diamantopoulos J, Stamou GP (1995) Analysis of the distribution of epiphytic li-
    chens within homogenous Fagus sylvatica stands along an altitudinal gradient (Mount Olym-
    pos, Greece). Vegetatio 116:33–40
Rai H, Khare R, Gupta RK, Upreti DK (2011) Terricolous lichens as indicator of anthropogenic
    disturbances in a high altitude grassland in Garhwal (Western Himalaya), India. Bot Orient
    8:16–23
Rai H, Upreti DK, Gupta RK (2012) Diversity and distribution of terricolous lichens as indicator
    of habitat heterogeneity and grazing induced trampling in a temperate-alpine shrub and mea-
    dow. Biodivers Conserv 21:97–113
Rawat R (2010) Uttarakhand, Simple relief map. http://uttarakhand.org/wp-content/up-
    loads/2010/08/map-relief.png. Accessed Feb 2013
Renner B, Galloway DJ (1982) Phycosymbiodemes in Pseudocyphellaria in New Zealand. Myco-
    taxon 16:197–231
Romeike J, Friedl T, Helms G, Ott S (2002) Genetic diversity of algal and fungal partners in four
    species of Umbilicaria along a transect of the Antarctic Peninsula. Mol Biol Evol 19:1209–1217
Sand-Jensen K, Jespersen TS (2012) Tolerance of the widespread cyanobacterium Nostoc com-
    mune to extreme temperature variations (−269 to 105 °C), pH and salt stress. Oecologia
    169(2):331–339
Santesson R (1952) Foliicolous lichens I. A revision of the taxonomy of the obligately foliicolous,
    lichenized fungi. Symb Botanicae Upsalienses 12:1–590
Scheidegger C, Clerc P (2002) Erdbewohnende Flechten der Schweiz, In: Rote Liste der gefährde-
    ten Arten der Schweiz: Baum- und erdbewohnende Flechten, p 75–108
Scheidegger C, Frey B, Zoller S (1995) Transplantation of symbiotic propagules and thallus frag-
    ments: methods for the conservation of threatened epiphytic lichen populations. Mitteilungen
    der Eidgenössischen Forschungsanstalt für Wald, Schnee und Landschaft 70:41–62
Sharma CM, Baduni NP, Gairola S, Ghildiyal SK, Suyal S (2010) Effects of slope aspects on forest
    compositions, community structures and soil properties in natural temperate forests of Garhwal
    Himalaya. J Forest Res 21:331–337
Singh JS, Singh SP (1987) Forest vegetation of the Himalaya. Bot Rev 53:80–192
Skaloud P, Peksa O (2010) Evolutionary inferences based on ITS rDNA and actin sequences reveal
    extensive diversity of the common lichen alga Asterochloris (Trebouxiophyceae, Chlorophy-
    ta). Molecular Phylogenetics Evol 54:36–46

Sundriyal RC (1992) Structure, productivity and energy flow in an alpine grassland in the Garhwal Himalaya. J Veg Sci 3:15–20

Tarhanen S, Holopainen T, Oksanen J (1997) Ultrastructural changes and electrolyte leakage from ozone fumigated epiphytic lichens. Ann Botany 80:611–621

Thüs H, Muggia L, Perez-Ortega S, Favero-Longo SE, Joneson S, O'Brien H, Nelsen MP, Duque-Thüs R, Grube M, Friedl T, Brodie J, Andrew CJ, Luecking R, Lutzoni F, Gueidan C (2011) Revisiting photobiont diversity in the lichen family Verrucariaceae (Ascomycota). Euro J Phycol 46:399–415

Tschermak-Woess E (1989) Developmental studies in trebouxioid algae and taxonomical consequences. Ibid. 164:161–195

Uher B, Aboal M, Kovacik L (2005) Epilithic and chasmoendolithic phycoflora of monuments and buildings in South-Eastern Spain. Cryptogamie, Algol. 24:275–358

Upreti DK (1998) Diversity of lichens in India. In: Agarwal SK, Kaushik JP, Kaul KK, Jain AK (eds) Perspectives in Environment, APH Publishing Corporation, New Delhi, India. p 71–79

Vishwakarma MP, Bhatt RP, Joshi S (2012) Macrofungal diversity in moist temperate forests of Garhwal Himalaya. Indian J Sci Technol 5:1928–1932

Voytsekhovich A, Kondratyuk S, Beck A (in press) Lichen photobionts of the rocky outcrops of Karadag Nature Reserve (South-East Crimea, Ukraine). J Phycol (in press).

Voytsekhovich AA (2013) Lichen photobionts: the origin, diversity and relationships with mycobiont.—Lambert Acdemic Publishing, Saarbrücken, 102 p IBSN: 978-3-659-31872-6. (In Russian)

Voytsekhovich AA, Kashevarov GP (2010) Pigment content of photosynthetic apparatus of green algae (Chlorophyta)—the photobionts of lichens. Int J Algae 12:271–281

Wang K, Sun J, Cheng G, Jiang H (2011) Effect of altitude on surface air temperature across the Qinghai-Tibet Plateau. J Mt Sci 8:808–816

Werth S, Tømmervik H, Elvebakk A (2005) Epiphytic macrolichen communities along regional gradients in northern Norway. J Veg Sci 16:199–208

Wetmore CM (1960) The lichen genus Nephroma in North and Middle America. Publications of the Museum. Michigan State University. Biol Ser 1:369–380

Will-Wolf S, Geiser LH, Neitlich P, Reis AH (2006) Forest lichens communities and environment—How consistent are relationships across scales? J Veg Sci 17:171–184

Zeitler I (1954) Untersuchungen über die Morphologie, Entwicklungsgeschichte und Systematik von Flechtengonidien. Österreichische botanische Zeitschrift 101:453–483

Glossary

Acicular Needle-shaped

Acuminate Gradually narrowing to a point, like a spade on a playing card

Acute Sharply pointed at the apex

Adnate Tightly adherent to surface

Aggregated Clustered

Amyloid Staining blue, blue-purple, blue-black or reddish in iodine

Anastomosing A joining together to form a vein-like network

Anisotomic Unequal branching, with a distinct main axis and smaller side branches

Alga (algae) A simple plant composed of a single cell or a string of cells

Apothecia A disk- or cup-shaped spore-producing organ

Apical Situated at the tip

Appressed Lying flat or pressed closely against the substrate

Arachnoid Cobweb-like in texture or pattern

Areolate Sharply divided into tile-like areolae

Areole (areolae) A small, irregular, often angular patch of thallus delimited by cracks or chinks in the thallus surface

Ascending Directed upwards at a rather narrow angle, curving upwards

Ascus (asci) The sac-like structure in which the spores are formed

Aseptate Simple without cross walls

Bacilliform Rod-like, usually more than 3 times as long as wide. cf. cylindrical

Biatorine Apothecia lacking a true exciple when mature and generally strongly convex

Biseriate In two rows

Bitunicate With two functional wall layers

Sporoblastidia Small subsidiary locules in a thick-walled spore—especially in *physciaceae*

Bullate Of a surface, blistered or puckered

Bullate-areolate With convex (blister-like) areolae

Canaliculated Channelled

Capitate Having a well-formed head

H. Rai, D. K. Upreti (eds.), *Terricolous Lichens in India,*
DOI 10.1007/978-1-4614-8736-4, © Springer Science+Business Media New York 2014

Cephalodium (pl. cephalodia) Delimited region within, or a warty, squamulose or shrubby structure on the surface of lichen thallus containing a photobiont (cyanobacterium) different to that characteristic of the rest of the thallus

Clavate Club-like

Concave Hollowed out, basin-like

Concolorous Of the same color throughout

Confluent Coming together; running into one another

Conglutinate Glued together

Conidioma (pl. conidiomata) Multi-hyphal, conidium containing structure

Constricted Of lobes, of varying width

Contiguous Touching but not fused

Convex Equally rounded, broadly obtuse

Convolute Of lobes, the upper surface strongly convex and the lower surface strongly concave

Coralloid Usually of isidia, coral-like, often brittle

Coriaceous Leathery

Cortex The outermost layer of the thallus, which if present, consists of hyphae which may appear either cellular or fibrous

Corymbose Arranged in clusters

Crateriform Crater-like

Crenate Having the edge toothed with rounded teeth

Crenulate Delicately crenate

Crustose Crust-like lichens that are closely attached to their substrate and lack a lower cortex

Crisped Of a margin, crumpled or thrown into waves

Cryptolecanorine Of an ascoma, with a reduced or inapparent thalline margin

Cuneate Wedge-shaped, thinner at one end than the other

Cyphella (cyphellae) A pore recessed into the lower thallus surface where medullary hyphae protrude

Cylindrical Rod-like, usually 2–3 times as long as wide. cf. bacilliform

Dactyl A hollow, nodular to cylindrical protuberance, somewhat resembling a swollen isidium, bounded by a cortex, often opening at the apex to expose the medulla

Decorticate Lacking a cortex

Decumbent Resting on a substratum, with the end turned up

Deflexed Bent sharply downwards

Delimited Having a distinct restricting edge or margin

Dendroid Irregularly branched, tree-like in form but not in size

Dichotomous Branching, frequently successively into two more or less equal arms; (cf. anisotomic, isotomic)

Diffract Cracked into small areas, areolate

Diffuse Widely or loosely spreading, with no distinct margin

Digitate Finger-like

Dimorphic Having two forms e.g., both a crustose and fruticose thallus as in *Cladonia*

Disc The upper surface of lichen apothecium enclosed by, but not including, the margin

Dorsiventral Flattened, with upper and lower surfaces

Ecorticate Without a cortex

Effigurate Obscurely lobed

Effuse Stretched out flat, especially as a spreading growth without a distinct margin

Ellipsoidal Oval in outline and three-dimensional

Endemic Occurring naturally only in a single geographic area

Entire Without teeth; more or less smooth on the margin

Epicortex A thin homogeneous polysaccharide-like layer over the surface of the cellular cortex which may have regular pores when viewed with the scanning electron microscopy

Epispore The fundamental and often outer wall of a spore which determines its shape (cf. perispore)

Epihymenium Uppermost (often pigmented) layer of hymenium, above asci

Epithecium The uppermost portion of the hymenium formed by the tips of the paraphyses, which are frequently expanded or branched, often pigmented, and sometimes containing tiny granules

Epruinose Lacking pruina

Erose Eroded

Erumpent Bursting through the surface

Evanescent Short-lived

Exciple The margin around the apothecial disc

Fasciculate Of branching or growth form, many branches arising from one point like a bundle of sticks; of rhizines, many simple rhizines arising from one point or region

Farinose Of soredia, like grains of flour (use × 10 lens)

Foveolate Honey-combed

Fenestrate With open areas or slits

Filamentous Thread-like

Filiform Very narrow in section

Fissured Cracked, split

Flabellate Fan-shaped

Flexuose Having a wavy or zigzag form

Foliose Having leaf-like lobes with distinct upper and lower surfaces

Fruticose A shrubby or hair-like growth form attached only at the base or free growing and normally with no clearly distinguishable upper and lower surfaces

Furcate Forked, as in rhizines with two long, terminal branches

Fusiform Spindle-like, narrower at the ends than in the middle

Glabrous Without indumentums

Glaucous Having a bluish grey bloom

Globose Globe-shaped

Granular Like grains of sugar

Gyrose Circularly folded or brain-like

Halonate Of the outer layer of spores, surrounded by a transparent coat

Hemiangiocarpic Of a sporocarp, opening before quite mature

Heteromerous Having mycobiont and photobiont components in well-defined layers, with the photobiont in a more or less distinct zone between the upper cortex and the medulla

Holdfast A process from the base of the thallus for attachment, often disc-like

Homoiomerous Having mycobiont and photobiont components intermixed throughout thallus, not layered

Hyaline Colorless, translucent

Hymenium The spore-bearing layer of fungal reproductive structures (ascoma)

Hyphae Fungal filaments, often modified and resembling round or angular cells

Hypothallus The first and purely fungal (without photobiont) layer upon which an algae-containing thallus may develop often projecting beyond the thallus onto substrate

Hypothecium The tissue just below the hymenium (and subhymenium) but above the exciple, often with a distinctive color or texture but sometimes merging with the exciple

Imbricate With overlapping layers

Immarginate Without a margin

Immersed Embedded in the substratum

Indeterminate Effuse

Innate Immersed

Involute With margins rolled in

Isidia Small, asexual reproductive structures on lichens that are minute and finger-like, covered with a cortex and that contain the photobiont

Isidiod Resembling isidia

Isidiate Having isidia

K Medullary reaction to potassium hydroxide (of chemical reactions)

KC Medullary reaction to potassium hydroxide followed by calcium hypochlorite

Labriform Lip-shaped

Lacerate With irregularly cut or torn margins

Laciniate Deeply, usually irregularly divided into narrow, more or less pointed segments; of lobes, developing laciniae or being lacinia-shaped; of margins, deeply, usually irregularly, divided into narrow, ± pointed segments

Laminal In the middle, or main part, of the thallus surface, rather than on the margins

Lateral At or near edge, especially side or secondary branches

Lax Loosely arranged

Lecanorine An apothecial margin which usually contains a photobiont and often resembles the thallus, but not the disk, in color and texture

Lecideine An apothecial margin with no photobiont cells that often resembles the disk, but not the thallus, in color and texture

Leprose Having the surface dissolved into soredia, loose, powdery, without any cortex

Lichen Composite organism made up of a fungus and an alga, a cyanobacterium, or all three

Linear Very narrow, with parallel margins

Lobate Bearing lobes

Lobe A flattened branch or projection

Lobulate Having lobules

Lobule Tiny, lobe-like, dorsiventral asexual reproductive outgrowths

Maculate Spotted or blotched

Margin Referring either to the outer edge of foliose or crustose lichen thalli or the outer boundary of apothecia

Marginate With a well-defined edge or margin

Matt With a dull surface

Medulla A loosely arranged layer of hyphae below the upper cortex and photobiont zone

Membranaceous Parchment-like

Monophyllous Consisting of a single lobe, often undulate or folded

Multiseptate With many septa

Muriform Like a wall, having many transverse and longitudinal septa. cf. submuriform

Mycobiont The fungal component of lichen

Oblong Proportioned about 1:3–6 with the margins more or less parallel; rectangular but ends not necessarily squared off

Oblique With sides unequal

Oblong Having the form of a rectangle of greater length than width

Obtuse Rounded or blunt at the apex

Ochraceous Of a dull, yellow color

Ocular chamber A cavity lying on the longitudinal axis of an ascus and penetrating into the thickened apical dome of the ascus from the ascal sac

Opaque Dull, not translucent

Orbicular Circular or nearly so, more or less flat

Oriented Turned in one direction

Ostiole An opening or pore, in fungi and lichens, a pore at the apex of a perithecium through which spores are extruded. adj. ostiolar

Oval Broadly elliptic, narrowing somewhat from middle to rounded ends

Ovoid Egg-shaped, three-dimensional

Ostiole A small opening or pore

P Medullary reaction to a fresh alcoholic solution of paraphenylene diamine (of chemical reactions

Palmate Radiately lobed or divided

Papilla (papillae) Minute protuberance on the surface of a cell

Paraphyses A sterile filament (sometimes branched, attached at the base and free at the summit) found amongst the asci in the hymenium

Parathecium Of apothecia, the outside hyphal layer

Pedicellate Stalked

Peltate Shield-like

Pendulous Hanging down from a support

Periclinal Curved in the direction of, or parallel to, the surface or the circumference

Perithecia A globose or flask-shaped fruiting body (ascoma) completely enclosed with protective sterile tissue and with an opening pore at the tip

Photobiont The photosynthetic component in lichen, either algae in the strict sense (e.g., green algae) or cyanobacteria (blue-green algae), or both

Polyphyllous Of a foliose thallus, divided into many lobes

Podetium (podetia) The upright, hollow stalk formed by an elongated apothecium

Proper exciple Tissue at the margin of an apothecium adjacent to the hymenium and hypothecium and inside the thalline exciple when present, without photobiont cells

Prosoplectenchyma Tissue consisting of cells with thickened walls and longish lumina and in which hyphal elements are recognizable as hyphae

Prothallus A weft of fungal hyphae (white, reddish or blue-black) at the margin of the thallus, lacking photobiont, often projecting beyond the thallus onto the substratum

Pruina Powdery frost-like deposit, typically composed of calcium oxalate

Pruinose Having a frosted appearance caused by a deposit of pruina

Pseudocyphella (pseudocyphellae) A break or opening in the cortex where medullary hyphae protrude; it may be round, irregular, angular, or a minuscule pore

Pubescent Having a somewhat dense cover of short, weak, soft hairs

Pulvinate In cushions

Punctiform Dot-like

Pustule A pimple or blister-like swelling, hollow within, often eroding. adj. postulate

Pycnidium (pycnidia) Minute, flask-shaped, fungal fruiting body

Pyriform Pear-shaped

Radiating Spreading from a central point

Reniform Kidney-shaped

Reticulum A network. adj. reticulate

Revolute Of a margin, rolled downwards; of lobes, weakly convolute, the upper surface weakly convex, the lower surface canaliculated

Recurved Curved downward or backward

Rhizine Root-like hyphae on the lower side of a foliose lichen thallus

Rimose Cracked

Rosette A circular cluster, e.g., of lobes

Rugose Wrinkled

Rugulose Delicately or minutely wrinkled

Rosette A flower-like pattern arrayed around a common point of attachment

Scabrous Rough to the touch with short, hard emergences or hairs

Schizidia Propagule formed from upper layers of thallus splitting off as scale-like segments from main lobes

Scrobiculate Coarsely pitted, foveolate

Scyphus An expanded, cup-like structure often terminating a podetium

Secondary metabolite Natural product of restricted taxonomic distribution with no obvious metabolic function

Septate Divided by cross walls

Septum A cross wall. pl. septa

Seriate Arranged in rows

Sessile Attached directly to the thallus surface without a stalk of any kind

Simple Not divided; unbranched

Soralium (soralia) An area of the thallus in which the cortex has broken down or cracked and soredia are produced

Soredium (soredia) Asexual reproductive structure that is powdery to granular, not covered with a well-defined cortex, and contains both algal (photobiont) and fungal (mycobiont) components

Sorediate Having soredia

Spores Microscopic reproductive bodies released from the apothecia of lichen

Spot test Tests for color reactions obtained by applying a liquid chemical reagent to lichen

Squamiform Scale-like

Squamule Small flakes or scales of lichen, lifting from the substrate, at least at the edges, often rounded, ear-like, or lobed

Squamulose Composed of or characterized by having squamules—an intermediate growth form between crustose and foliose

Squarrose Branching at right angles from a single main axis, like a bottlebrush

Sterile Without sexual reproductive structures

Subfoliose Almost foliose, pertaining to the overall growth form of a crustose thallus that has marginal lobes showing some tendency to curve upwards

Submuriform Of spores, having both transverse and longitudinal septa, but in which not more than 15 cells may be seen. cf. muriform

Substratum The underlying layer or base to which lichen is fixed

Subulate Tapering from a wide base to a sharp apex, more or less circular in cross section, awl-shaped

Superficial On the surface

Terete Circular in cross-section—cylindrical and smooth

Terminal Borne at the end

Thalline exciple Tissue at the margin of an apothecium external to proper exciple and having a structure similar to that of the vegetative thallus with photobiont cells included in it

Thallus The vegetative part of lichen, a more or less undifferentiated plant body

Tholus A thickened inner part of the ascus wall in the ascus apex

Tomentose Densely covered with matted short hairs

Thallus (thalli) The vegetative body consisting of both algal and fungal components, not differentiated into a stem and leaves

Translucent More or less transparent

Transverse Across the width

Trichotomous Branching almost equally in three parts

Truncate With an abruptly transverse end, as if cut off

Tufted Of rhizines, a simple rhizine densely fasciculate at the tip

TLC Thin layer chromatography—a technique used to separate chemical compounds

Tomentum A layer of hair-like structures other than discrete rhizines

UV Response of cortex to UV light

Index

H. Rai, D. K. Upreti (eds.), *Terricolous Lichens in India,*
DOI 10.1007/978-1-4614-8736-4, © Springer Science+Business Media New York 2014

Printed by Publishers' Graphics LLC